Structural Change and Performance of the U.S. Grain Marketing System

Structural Change and Performance of the U.S. Grain Marketing System

Edited by

Donald W. Larson
Department of Agricultural, Environmental, and Development Economics
The Ohio State University

Paul W. Gallagher
Department of Economics
Iowa State University

Reynold P. Dahl
Department of Applied Economics
University of Minnesota

Produced by

Chris Scherer
Scherer Communications
Urbana, Illinois

Donald W. Larson, ed
Paul W. Gallagher, ed
Reynold P. Dahl, ed

Structural Change and Performance of the U.S. Grain Marketing System

Production Coordinator: Chris Scherer, Scherer Communications, Urbana, IL 61801
Technical Editor: Emily Van Hyning, Champaign, IL 61820
Designer: Gretchen Wieshuber, Champaign, IL 61820
Printer: Brian Walker, Prestige Printing, Champaign, IL 61820

Library of Congress Catalog Card Number: 98-61153

ISBN 0-9647738-4-8

Acknowledgments

Publication of this book was supported by the Departments of Agricultural Economics and Agricultural Experiment Stations at the: University of Arkansas, University of Georgia, University of Idaho, University of Illinois, Iowa State University, Kansas State University, Louisiana State University, University of Minnesota, Mississippi State University, University of Nebraska, North Dakota State University, The Ohio State University, Oklahoma State University, Oregon State University, Purdue University, and the U.S. Department of Agriculture.

We acknowledge the contributions of the following authors: **David W. Bullock**, chief economist, Minneapolis Grain Exchange; **Patricia Callow (Carlson)**, former research assistant, Department of Agricultural Economics and Rural Sociology, University of Idaho; **Reynold P. Dahl**, professor emeritus, Department of Applied Economics, University of Minnesota; **Bruce Dahl**, research scientist, Department of Agricultural Economics, North Dakota State University; **Paul L. Farris**, professor emeritus, Department of Agricultural Economics, Purdue University; **Paul W. Gallagher**, associate professor, Department of Economics, Iowa State University; **Wayne M. Gauthier**, associate professor, Department of Agricultural Economics and Agricultural Business, Louisiana State University; **Lowell D. Hill**, L. J. Norton Professor of Agricultural Marketing, Department of Agricultural and Consumer Economics, University of Illinois at Urbana-Champaign; **Jack E. Houston**, associate professor, Department of Agricultural and Applied Economics, The University of Georgia; **Bob F. Jones**, professor emeritus, Department of Agricultural Economics, Purdue University; **James R. Jones**, professor, Department of Agricultural Economics and Rural Sociology, University of Idaho; **George W. Ladd**, professor emeritus, Department of Economics, Iowa State

University; **Donald W. Larson**, professor, Department of Agricultural, Environmental and Development Economics, The Ohio State University; **Robert L. Oehrtman**, professor, Department of Agricultural Economics, Oklahoma State University; **Lu Qu**, former research assistant, Department of Agricultural Economics and Rural Sociology, University of Idaho; **Eric J. Wailes**, professor, Department of Agricultural Economics and Agribusiness, University of Arkansas, Fayetteville; **Maurice V. Wiese**, professor, Department of Plant, Soil, and Entomological Sciences, University of Idaho; and **William W. Wilson**, professor, Department of Agricultural Economics, North Dakota State University

In Honor of Reynold "Reese" P. Dahl

Over the past several decades, agricultural economists have made great strides in understanding agricultural markets, agricultural marketing and transportation. In turn, they've made significant contributions to informed public policy and private decision making. Over this period, Reynold P. "Reese" Dahl has been a leader in exploring the complexities of agricultural markets and in relating his findings to the real world of agriculture. It is fitting then to dedicate this book to him.

Reese has been a mainstay in the North Central region and in the nation in research and education related to agricultural marketing. Even though he was promoted to the rank of professor emeritus in 1994, he has remained engaged and curious about actions, policies, and the factors which influence the spatial and temporal dimensions of agricultural markets.

Reese's impact on agriculture and agricultural economics reaches well beyond his own work. In every way he has been a mentor and collaborator. Those of us fortunate enough to have had a close association with Reese have benefited from both his insights and his unfettered enthusiasm. And as one myself, I can attest to his intense loyalty to and unending interest in his students. Once you've been Reese's student, you're always his student.

For years Reese has hosted regional meetings, organized seminars and done those things that bring people together to discuss important issues. Because he's never a reticent participant in discussions, others are drawn in. Reese challenges, probes, and opines in the most positive spirit of intellectual debate. His infectious enthusiasm for the work of agricultural economists has inspired and encouraged those who produce books such as this one.

Reese Dahl and his generation of professionals have created a foundation on which a wide and deep literature has been built. No tribute could be greater than to have students gain new perspectives about agricultural markets and marketing from this book.

Michael V. Martin
Dean, College of Agricultural, Food, and Environmental Sciences
Acting Vice President for Agricultural Policy

Foreword

This book is a product of the cooperative efforts of the North Central Regional Grain Marketing Research Committee (formerly NC186) whose members are faculty from several major United States universities. The book consists of papers presented at numerous workshops and conferences organized by the Committee from 1992 to 1997. Without the active participation and support of the regional research committee members, the book would not have been published. As editors we wish to thank all of the committee members for their valuable contributions to this effort. The book is designed to be a comprehensive study of structural change and performance of the U.S. grain marketing system in the last 20 years.

The book chapter tells a story of boom and bust in world and U.S. grain markets that have caused a great deal of structural change in grain marketing. Firms expanded through mergers and acquisitions in the good times, and later in the bad times, were forced to reduce excess capacity to become more competitive. Firms also had to seek new value added markets to increase profit margins. Although the adjustments may have been difficult, the net result is that the grain marketing system may be much better positioned for the future because of these adjustments.

The first seven chapters describe the changing grain marketing system, address issues of trade and export expansion and decline, discuss the changing role of futures and options markets, analyze the effects of a changing transportation system on the grain marketing system, describe organizational change in grain industry, as well as issues of public grain grades and phytosanitary standards in wheat exports.

The next seven chapters examine the dynamics of changing structure in the flour milling, wet corn milling, soybean processing, oilseed processing, rice milling, malting barley, and feed manufacturing industries.

The last chapter contains the conclusions and implications for the future of the grain industry.

We hope the book will be of great value to grain trading and processing firms, grain transportation firms, commodity associations, researchers, students and others who want to learn more about our grain marketing system.

Donald W. Larson, Paul W. Gallagher, and Reynold P. Dahl, editors

Contents

1 THE CHANGING GRAIN MARKETING SYSTEM 1
Reynold P. Dahl

2 U.S. GRAIN EXPORTS: GROWTH AND DECLINE. 7
Bob F. Jones

3 FUTURES, OPTIONS, AND CASH MARKETS 25
William W. Wilson, David W. Bullock, and Bruce Dahl

4 GRAIN TRANSPORTATION AND GRAIN INDUSTRY
STRUCTURE. 51
George W. Ladd

5 STRUCTURAL CHANGE IN THE GRAIN MARKETING
INDUSTRY 109
Reynold P. Dahl

6 THE CASE FOR PUBLIC GRAIN GRADES 135
Lowell D. Hill

7 PHYTOSANITARY STANDARDS: THE CASE OF PACIFIC
NORTHWEST WHEAT SHIPMENTS TO CHINA 151
James R. Jones, Patricia Carlson, Lu Qu and Maurice V. Wiese

8 STRUCTURAL CHANGES IN THE NORTH AMERICAN
FLOUR MILLING INDUSTRY 165
William W. Wilson

9 GROWTH AND STRUCTURE OF THE WET CORN
MILLING INDUSTRY 181
Paul L. Farris

10 U. S. SOYBEAN PROCESSING INDUSTRY: STRUCTURE
AND OWNERSHIP CHANGES 201
Donald W. Larson

11 STRUCTURAL CHANGE IN THE U.S. EDIBLE VEGETABLE
OILS PROCESSING INDUSTRY. 217
Jack E. Houston

12 U.S. RICE MILLING INDUSTRY: STRUCTURE AND
OWNERSHIP CHANGES 233
Eric J. Wailes and Wayne M. Gauthier

13 NORTH AMERICAN MALTING INDUSTRY. 253
William W. Wilson

14 AMERICAN FEED MANUFACTURING INDUSTRY:
TRENDS AND ISSUES 273
Jack E. Houston

15 SUMMARY AND CONCLUSIONS 287
Bob F. Jones and Robert L. Oehrtman

Figures

Figure 1.1   U.S. Grain Exports and Ending Stocks 3

Figure 2.1   United States Share of World Grain Exports 8

Figure 2.2   Total Soybean Exports 8

Figure 2.3   Total Grain Exports 10

Figure 2.4   Total Grain Imports (former Soviet Union
Eastern Europe) 14

Figure 2.5   U.S. Grain Storage Capacity 16

Figure 2.6   EC-12, Net Grain Trade 21

Figure 3.1   Average Daily Volume for Selected Grain Futures,
by Year 28

Figure 3.2   Average Daily Volume for Selected Livestock Futures,
by Year 29

Figure 3.3   Average Daily Volume for Selected Commodity/Index
Futures, by Year 30

Figure 3.4   Average Daily Open Interest for Selected Livestock
Futures, by Year 31

Figure 3.5   Average Daily Open Interest for Selected Grain Futures,
by Year 32

Figure 3.6   Average Daily Open Interest for Selected Commodity/
Index Futures, by Year 33

Figure 4.1   Average Cost Exceeds Marginal Cost 55

Figure 4.2   Cost of Flotation Related to Size of Loan 59

Figure 4.3   Average Total Cost with and without Shared Cost 60

Figure 4.4   Average Total Cost Curves 65

Figure 5.1   Grain Marketing, Distribution Channels, and Methods
of Transportation. 110

Figure 7.1   Costs to Deliver Wheat to China from North Idaho . 155

Figure 8.1   Market Share for Grain Companies and Vertical
Integrated Firms: United States 169

Figure 8.2   United States Flour Milling 4-Firm Market Share and
Herfindahl Index 170

Figure 8.3   Herfindahl Index for United States Flour Milling
Industry, by Census Bureau Regions for 1972
and 1992 171

Figure 8.4   Number of Flour Milling Plants Closed, Merged,
and Opened for Canada and United States 172

Figure 10.1   U.S. Soybean Processing Plants, 1994. 204

Figure 12.1   1994 Location of U.S. Rice Mills 238

Figure 12.2   Marketing Margins of Arkansas Rice Mills 244

Figure 12.3   Marketing Margins of Houston Rice Mills 245

Figure 12.4   Marketing Margins of California Rice Mills 246

Figure 12.5   U.S. Rice Farm-Mill Price Margins, (milled basis,
1982$), 1978-90 247

Figure 13.1   Beer Production by State 259

Figure 13.2   North American Malting Plants and Barley
Production 261

Tables

Table 3.1   Percentage Change in Average Daily Volumes and Open Interest Between 1980-90 and 1990-97 for Selected Futures Contracts. ... 27

Table 4.1   Relations Between Marginal and Incremental and Demand Elasticity: e_{ij} and v_{ji} $\pi 0$. ... 84

Table 5.1   Futures Contracts Traded on U.S. Grain Futures Markets, by Commodity, Selected Years. ... 115

Table 5.2   Ten Largest U.S. Multiple Facility Grain Companies by Number of Facilities and Capacity, 1995. ... 118

Table 5.3   Ten Largest U.S. Grain Elevator Companies, 1981. ... 120

Table 5.4   Joint Ventures in Grain Marketing and processing, 1995. ... 124

Table 5.5   Grain Storage Capacity in the U.S., On-Farm and Off-Farm, by State, 1 April 1987 and 1 December 1993. ... 128

Table 7.1   China's TCK Quarantine and Pacific Northwest Wheat Shipments (1,000 metric tons) and Prices ($ per ton). ... 161

Table 8.1   Descriptive Flour Industry Milling Statistics. ... 168

Table 8.2   Plant Characteristics: Exited, New, and Merged (1972-1990). ... 173

Table 8.3   Correlation of Acquired and Acquiring Firm Characteristics: United States (N=80). ... 174

Table 8.4   Productivity Comparisons in Milling Industries Between Mexico and the United States. ... 178

Table 9.1   Corn: Food, Seed and Industrial Use, 1980-81 to 1996-97. 183

Table 9.2   HFCS and Sugar Prices and Producer Price Index for Total Finished Goods, 1981-94. 184

Table 9.3   Corn Product Shipments by the Corn Wet Milling Industry, (Census Industry 2046). 185

Table 9.4   Corn Sweetener Consumption Per Capita in the United States, 1991-96. 186

Table 9.5   U.S. Domestic Food and Beverage Uses of HFCS-42 and HFCS-55, 1980 and 1995. 187

Table 9.6   World Production of High Fructose Corn Syrup for Selected Countries, Selected Years. 187

Table 9.7   Percentage of Total Value of Shipments Accounted for by Largest Companies in the Corn Wet Milling Industry (SIC 2046), Selected Years. 191

Table 9.8   U.S. production Capacity for High Fructose Corn Syrup, by Company, 1987 and 1992. 192

Table 9.9   Corn Price and Cost to the Corn Wet Milling Industry, 1982, 1987, and 1992. 195

Table 9.10   U.S. High Fructose Corn Syrup (HFCS) Supply and Use, 1990-96. 197

Table 10.1   Soybean Processors in the U.S. and Number of Plants, 1994. 205

Table 10.2   Top Four Firms and Four-Firm Concentration Ratio of Soybean Crushing Industry, With and Without Acquisitions 1977, 1982, 1988, and 1995. 206

Table 10.3   Estimated U.S. Soybean Processing Capacity, Number of Plants and Number of Firms, Selected Years. 207

Table 10.4   Number of Firms, Plants and Capital Expenditures by Soybean Mills, Selected Years 1954-1992. 209

Table 10.5   Distribution of Soybean Oil Mill by Number of Employees. 210

Table 10.6   Crush, Value of Shipments, Value Added by Manufacture, and Crushing Margins in the U.S. Soybean Processing Industry, 1964-94. 212

Table 10.7   Soybean Crush and Crush Margin Regression Results, 1964-94. 213

Table 11.1   World Food Use Consumption of Major Edible Vegetable Oils, 1979-91. 221

Table 11.2   U.S. Imports of Palm and Coconut Oils, 1979-91. . . . 223

Table 11.3   Estimated Own-Price, Cross-Price, and Income Elasticities of Soybean, Cottonseed, Palm, and Coconut Edible Oil Demand in the Ante- and Post-Campaign Period. 227

Table 11.4   Mean Estimated Own- and Cross-Price Elasticities and g for Soybean, Cottonseed, Palm, and Coconut Oil Demand in the Ante- and Post-Campaign Periods. 229

Table 12.1   Number of Rice Mills by State. 235

Table 12.2   Milling and Distribution Shares to U.S. Consumers of Direct Food Rice, Selected Years. 236

Table 12.3   Volume of Rice Milled, Farm and Mill Value, and Value Added, U.S., 1970-95. 240

Table 12.4   Rice Mill Volume, Value Added, and Milling Margin Trend Regression Results, 1970-95. 241

Table 12.5   Average Cost of Rice Milling, 1986. 242

Table 12.6   Cost Comparison Among Six Mill Centers. 243

Table 12.7   Regression Results of a U.S. Rice Farm-Mill Markup Pricing (MP) Model. 249

Table 13.1   U.S. Production Estimates of Malting Barley. 256

Table 13.2   Structural Characteristics of the U.S. Brewing Industry. 257

Table 13.3   Structural Characteristics of the North American Malting Industry. 260

Table 13.4   Comparative Statistics: U.S. and Canadian Malting. . . 264

Table 13.5   Summary of EEP in U.S. Barley Sector (1985-86 to 1991-92). 266

Table 14.1   Prepared Feeds (2048), N.E.C. (animals and fouls, except dogs and cats; includes slaughter for feeds). Employment and Product Value, 1972-1992. 275

Table 14.2   Top Feed Manufacturers in United States (1989). . . . 276

Table 14.3   Prepared Feeds (2048), N.E.C. (animals and fouls, except dogs and cats; includes slaughter for feeds). Employment and Product Value, 1992. 278

Table 14.4   Soybean Oil Crushing Industry (2075), Employment and Product Value, 1972-1992. 279

Table 14.5   Largest Four Firms, Soybean Oil Crushing Industry, 1977-1988. 280

Table 14.6   Wet Corn Milling Industry (2046). Employment and product Value, 1972-1992. 281

Table 14.7   Largest Four Firms, Corn Milling Industry, 1977-1988. 283

Table 14.8   Cottonseed Oil Industry (2074), Employment and product Value, 1972-1992. 284

Table 14.9   Largest Four Firms, Cottonseed Milling Industry, 1977-1988. 285

1

The Changing Grain Marketing System

Reynold P. Dahl

The U.S. grain marketing system is the vehicle by which huge quantities of grain valued at billions of dollars are moved each year from American farms to consumers in this and other countries. Grains and oilseeds form the base of the American food system, and the United States is the largest grain exporting country in the world. In calendar year 1994, exports of grains and feeds, oilseeds and products, totaled $20.7 billion, 45 percent of total U.S. agricultural exports.

Coordination of these large shipments, delivering the correct types and grades of grain when and where they are needed, is not an easy task. Yet it is accomplished with a decentralized free-market system in a remarkably efficient manner. The U.S. grain marketing system is a private enterprise system in which individual firms own the facilities and reap the rewards as well as the consequences of their own decisions. It is dynamic and changes in response to market forces. This is an important strength of a private enterprise system in contrast to government-owned and operated grain marketing systems that characterize many countries.

But changes in demand placed upon the system resulting from changes in economic variables such as grain production, exports, transportation, and government programs are frequently abrupt and difficult to predict. Hence, investments in marketing infrastructure are often risky and sometimes result in painful losses. The grain marketing system can move from undercapacity to excess capacity in a short time span. This induces structural change in the system.

Structural changes in the U.S. grain marketing system have been more extensive in recent years than at any time in history. The late 1980s and early 1990s can best be characterized as a period of consolidation and increased concentration in grain marketing. To understand the economics of these changes one has to look at the stimulus to investment in marketing infrastructure resulting from the grain export boom of the 1970s.

The 1970s will go down in history as the golden decade for American agriculture and its grain marketing system. After more than 25 years when surplus stocks and government price support operations dominated grain markets and marketing, the 1972-73 marketing year ushered in a new era. U.S. grain exports more than tripled in the 1970s reaching an all-time record of nearly 5.0 billion bushels in 1980 (Figure 1.1). The U.S. grain marketing system deserves considerable credit for accommodating this big increase in volume with a minimum of disruptions. Marketing margins increased as the demand for marketing infrastructure exceeded the available supply. A euphoria prevailed in the industry as a continued rapid growth potential was perceived. This stimulated investments in rail cars, barges, storage, and port facilities, much of which did not come on line until the 1980s when grain exports began an extended period of decline.

U.S. grain exports declined to 3,017 million bushels in 1986 from their record high of 4,951 million bushels in 1980. Competition for the reduced volume drove marketing margins down. The new investments in rail cars, unit-train grain loading, barges, and port elevators resulted in a surplus of such marketing infrastructure that became burdensome. U.S. grain exports increased to 4,536 million bushels in 1989, but declined to 3,449 million bushels in 1994, 1,502 million bushels below the 1980 record.

As grain exports declined in the 1980s, carryover grain stocks accumulated under federal farm programs, reaching an all-time high of 8.4 billion bushels at the end of the 1986-87 marketing year. The resulting increase in grain storage income in the grain industry offset, in part at

Figure 1.1 • U.S. Grain Exports and Ending Stocks

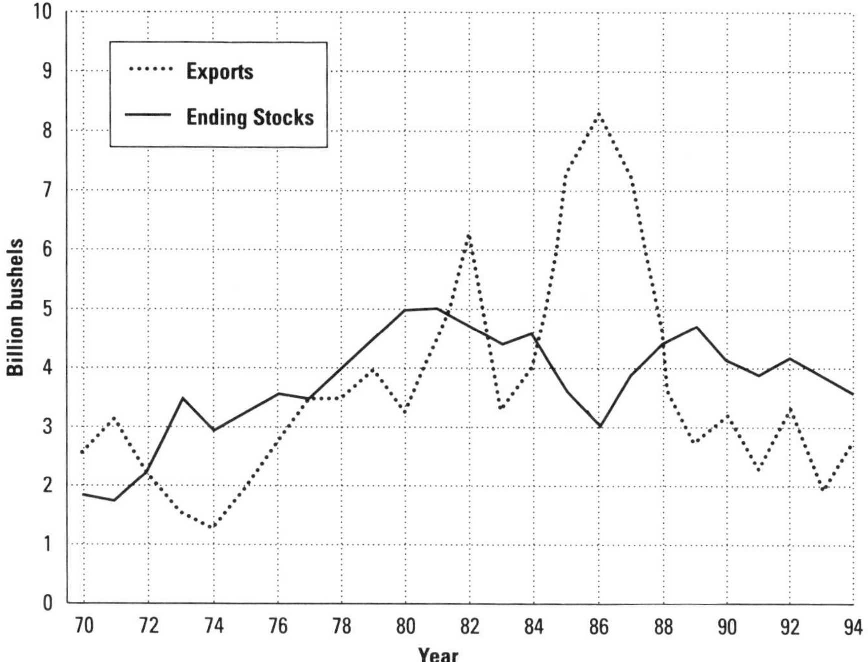

least, the decline in merchandising income associated with lower grain exports. But, stocks dropped precipitously following the drought in 1988, reaching 1.9 billion bushels at the end of the 1993-94 marketing year, their lowest level since 1975-76. This reduced grain storage income in the industry. Stocks rose to 2.7 billion bushels following the record 1994 corn crop, but are still low relative to past norms.

The reduction in both grain exports and grain stocks as described above has resulted in excess capacity and reduced marketing margins in the grain marketing industry. These along with other changes such as railroad deregulation and government programs have induced many structural adjustments.

Purpose and Organization

The purpose of this book is to describe and analyze structural changes in the U.S. grain marketing system and their performance implications. The economics of the grain export boom of the 1970s and the subsequent decline in grain exports in the 1980s are analyzed in Chapter 2. Changes in world grain production, world economic growth, monetary exchange rates, government agricultural programs to enhance farm incomes, and the Export Enhancement Program (EEP) are also discussed.

The impact of changes in grain exports and price variability on futures trading and hedging is discussed in Chapter 3, along with the increased importance of futures prices as a basis for cash grain prices. Options trading on grain and oilseed futures contracts emerged in the early 1980s. Options on grain futures have received increased volume of trade providing alternative hedging vehicles in the management of price risk. Options on futures and their role in risk management are analyzed in Chapter 3.

Grain and oilseeds are bulky commodities relative to their value. Hence, the cost of transportation is a major cost in grain marketing. Changes in grain transportation costs as impacted by railroad deregulation and intermodal competition have induced many structural changes in the grain marketing system. The effects of technology changes in transportation on optimal plant size are analyzed in Chapter 4.

Discussion of the impact on grain marketing channels of changing transportation costs associated with unit train rates and river barges is continued in Chapter 5. Their effects on the changing structure of country elevators, terminal elevators, grain exchanges, and the decentralization of cash grain trade are also discussed. Changes in the nation's largest multiple facility grain firms from 1980 to 1995 are also discussed in Chapter 5. Structural changes in grain marketing cooperatives at the local, regional, and interregional levels as well as structural change in the U.S. grain export system are analyzed in Chapter 5.

Joint ventures have become increasingly popular in grain marketing. Joint ventures are identified and described in each of two groups: (1) joint ventures between farmer-owned cooperatives and investor-oriented firms (IOF's), and (2) joint ventures between IOF's. Some possible advantages and disadvantages of joint ventures are discussed in Chapter 5. This chapter concludes with a discussion of increased vertical integration in the grain marketing system. Grain merchandising, exporting, and processing are being integrated along with more investments in value-added grain processing by both farmer-owned cooperatives and IOF's.

As U.S. grain exports declined in the 1980s, the alleged poor quality of U.S. grain was advanced as a possible explanation. This culminated with the passage of the U.S. Grain Quality Improvement Act of 1986. Grain grades and the quality of U.S. grain exports are analyzed in Chapter 6. A study of the Chinese quarantine of wheat exports through the Pacific Northwest is discussed in Chapter 7.

The U.S. grain marketing system also includes value-added grain processing industries where sizeable expansion and increased investments have occurred in recent years. Structural changes in flour milling, wet corn milling, soybean processing, oilseed processing, rice milling, malting barley, and feed manufacturing are analyzed in Chapters 8 through 14.

The final chapter presents the conclusions of the book and implications for the future growth of the grain industry.

2

U.S. Grain Exports: Growth and Decline

Bob F. Jones

Introduction

World import demand for grain and oilseeds experienced slow growth from 1960 through 1971. Annual U.S. exports of grains varied between 24 and 45 million tons. From 1960 through 1971, U.S. exports did not exceed 39 million tons except during the three-year export boomlet centered around 1965–66 when grain production in Southeast Asia was sharply reduced. The U.S. export market share for all grains was 34 percent in 1971 (Figure 2.1).

U.S. soybean exports expanded from 5.8 million tons in 1954 to 11.8 million tons in 1971. In 1971 the U.S. market share for soybeans was 94 percent (Figure 2.2).

Through the 1960–1971 period, U.S. agriculture was characterized as having excess production capacity. Non-resource loans and subsidized storage programs played a key role in price and income policy. In the early part of the period, ending stocks as a percent of total use were as

Figure 2.1 • United States Share of World Grain Exports

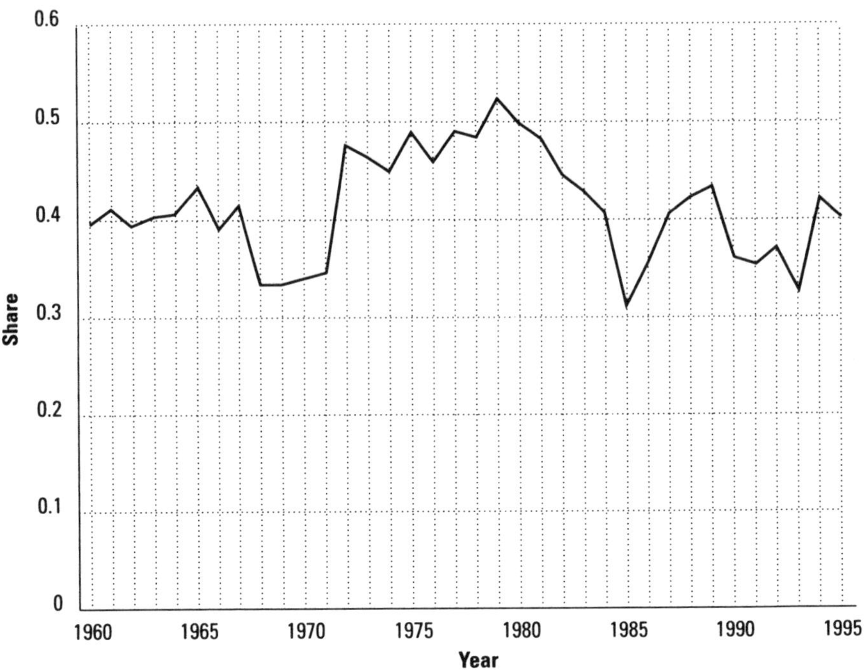

Figure 2.2 • Total Soybean Exports

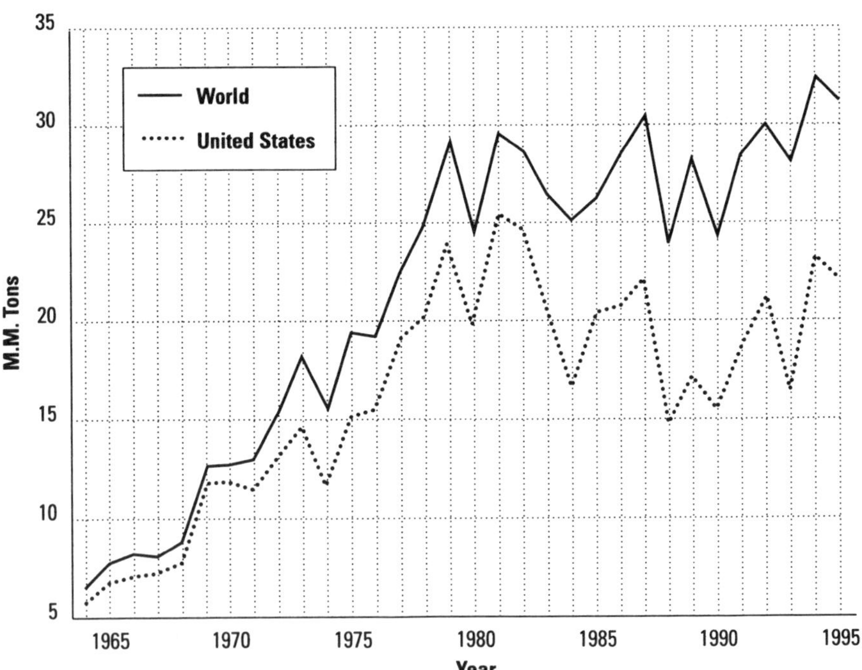

high as 130 percent for wheat, 55 percent for corn and about 35 percent for soybeans. A high proportion of these stocks was owned by the Commodity Credit Corporation (CCC) or were owned by producers who had placed the grain under regular or extended loan programs with credit extended by the CCC. Storage payments by the CCC were an important source of income for the grain marketing system. Also, the CCC provided subsidized credit for construction of grain storage facilities, both on-farm and off-farm.

During this period, a significant proportion of exported U.S. government-owned grain stocks was shipped under programs authorized by the Agricultural Trade Development and Assistance Act of 1954, often called P.L. 480. In the mid- to late 1950s, P.L. 480 shipments comprised about 40 percent of U.S. grain exports. That share declined in the 1960s as commercial exports expanded in part through use of direct export subsidy payments made to exporters.

In 1972 a series of world events began to unfold which had a dramatic effect on export demand, the U.S. grain marketing system, world grain prices, farm income, and consumer food prices. Some of these events had their roots in the previous two decades. In 1972 the volume of U.S. grain exports jumped from 37 million to 66 million tons, up almost 77 percent from the previous year. After 1972 U.S. grain exports continued to grow, reaching a peak of 114 million tons in 1980. The U.S. share of the world market for exports increased from 34 percent to 42 percent in one year. Soybean exports increased from 13 million tons to 14.7 million tons. However, with rapid growth in exports of Brazilian soybeans, the U.S. share slipped to 81 percent of world exports. By 1981, the peak year for U.S. soybean trade, exports had reached the 25 million ton level.

The purposes of this chapter are to: (1) identify the principal developments contributing to growth in the export market; (2) discuss the U.S. response to that growth; (3) identify subsequent developments which contributed to the decline in U.S. exports; and (4) briefly discuss longer-run prospects for U.S. export growth.

The Export Boom from 1972–1980: Principal Developments

A combination of five worldwide developments was the impetus for the export boom: (1) a decline in world grain production; (2) a shift in exchange rate policy; (3) continuing world economic growth; (4) a dramatic increase in petroleum prices; and (5) price and import policies followed in centrally planned economies. Although there is agreement that each of these developments contributed to the boom, there is not complete agreement on the relative weights to assign to each one. Each event and its implications for trade is discussed briefly.

Decline in World Grain Production in 1972 and 1974

World grain production declined 3.1 percent in 1972 from the 1971 level. On a worldwide basis, this is not a large change, but the decline was concentrated in four countries which were important in world grain trade: the Soviet Union, Australia, Brazil, and South Africa. By drawing down stocks, it was possible for world consumption to increase 2.0 percent in the same year that production declined. World trade in grain increased from 123 million tons in 1971 to 149 million tons in 1972 or 21.6 percent in one year (Figure 2.3). As noted above, U.S. grain exports increased about 77 percent from 1971 to 1972. The United States was able to respond to the sharp increase in demand because it had large amounts of grain in storage.

Figure 2.3 • Total Grain Exports

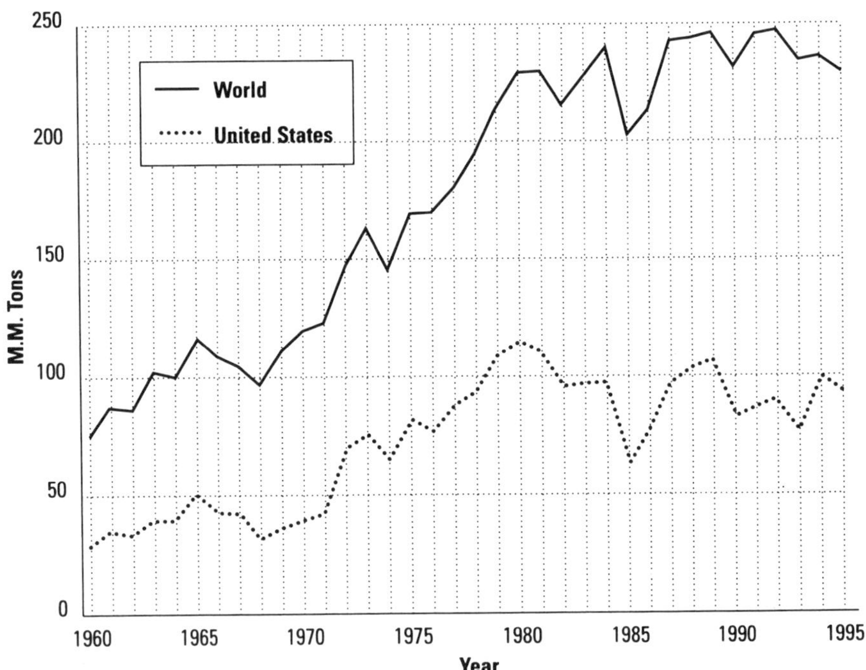

As a result of adverse weather conditions in the Soviet Union in 1972, their wheat crop was expected to be sharply reduced. (When the harvest was in, total grain production was down 12.8 million tons from the previous year, or 7.4 percent, with all of the decrease due to a decline in wheat production.) Prior to 1972, the Soviet Union had embarked on a five-year plan to increase food production, especially livestock production. The plan called for importing grain in order to maintain livestock herds if domestic production was reduced. In earlier periods of crop shortfalls, the Soviet Union had reduced livestock numbers when adverse weather limited grain production. With the sharply reduced crop, they implemented the plan and sought grain from the world market. Since the United States held large stocks of grain in government storage and was eager—in retrospect, probably too eager—to dispose of its surplus stocks, a bargain was struck. The United States would supply wheat to the Soviet Union at the prevailing world trading price of $1.65 per bushel. Since prices were being supported for U.S. producers above this level, export subsidies were required to move the wheat. During mid-1972, Soviet grain buyers agreed to purchase over 400 million bushels of wheat from the United States, at the time the largest single purchase of grain by one country. A purchase of this magnitude enabled the CCC to unload its stock of grain so that by the end of 1974 the CCC held no wheat under its control.

Australia, generally the third largest wheat exporter after the United States and Canada, experienced a severe drought in 1972 which reduced its production by about 23 percent. Consequently, Australian exports declined from about a 10 percent share of world wheat exports in 1970 and 1971 to less than 4 percent in 1972.

Brazil relies on the import market for a large share of its wheat consumption, at that time producing only about 50 percent of its annual needs. In 1972 the Brazilian wheat crop was down by 1.34 million tons or 66 percent from the previous year. With a population of over 100 million, the decreased domestic production generated a significant increase in Brazilian wheat imports.

South Africa, because of wide swings in weather, has a history of wide swings in its exports. It exports some corn and generally imports from 0.5 to 0.8 million tons of wheat every year. In 1972 South Africa grain crops were down 1.6 million tons or 21 percent from the previous year. Consequently, its exports of corn dropped from 1 million tons the previous year to 0.5 million tons.

Adverse growing conditions reduced world grain production again in 1974. World grain production fell 4.0 percent in 1974 from the 1973 level. This time the largest reduction was in the United States. U.S. corn production was reduced by about 17 percent from the previous year. Following closely on the 1972 crop shortfall and subsequent draw-down in stocks, world consumption declined by 3.8 percent. Likewise, world trade in grain declined. U.S. exports of grain declined 11.8 percent from the previous year.

Monetary Exchange Rates

From the Bretton Woods Conference in 1944, the United States and its major trading partners operated until 1973 with fixed exchange rates for each currency. Under this system the U.S. dollar had an important role as a major reserve currency. As a supplier of a major reserve currency, the United States facilitated the expansion of trade by running a persistent deficit in the current account of its balance of payments.

This system worked reasonably well until inflation accelerated in the United States in the late 1960s. Gradually the U.S. dollar became overvalued at fixed exchange rates relative to other currencies. Foreigners became less willing to hold U.S. dollars at established prices.

In a 13-month period in 1972 and 1973, the United States moved to a market-determined exchange rate. With persistent high rates of inflation in the United States during the 1970s, the dollar declined in value relative to other major world currencies. Rates fluctuated relative to the German mark and Japanese yen during the 1970s. Although the rate was fairly stable between 1973 and 1976, it fell about 17 percent between 1976 and 1980.

A declining exchange rate contributes to higher U.S. grain prices, *ceteris paribus*. Since a devaluation of the dollar means an upward valuation of the other currency, holders of the other currency have to put up fewer units of their currency to buy a unit of U.S. grain. At lower prices, foreigners tend to purchase more imported grain. Larger purchases lead to higher prices in dollars, especially if the supply of grain in the United States is relatively inelastic.

World Economic Growth

Most countries around the world experienced significant economic growth in the decades of the 1960s and 1970s. Although there were several short-term downturns over the two decades, the general trend was upward.

Rising incomes, especially in less developed countries, along with growing populations, generate increased demand for agricultural products. Many consumers add more livestock products to their diets if they can afford them. Increased demand for livestock products translates into growth in demand for feedgrains and oilseed meals. When domestic production is not expanding sufficiently to meet growing domestic demand for these products, many countries turn to export suppliers for these products. As long as real per capita income is rising around the world, these developments contribute significantly to growth in demand for U.S. feedgrains and oilseeds.

World Oil Prices

In a series of moves starting in 1973 the Organization of Petroleum Exporting Countries (OPEC) raised the price of crude oil from $3.01 to $11.65 per barrel. The Arab–Israeli War and the Arab oil embargo dramatized to the world the dependence of many countries on imported oil. Since importing countries continued to buy oil at these prices, and were either unable or unwilling to get OPEC to lower its prices, the era of cheap energy came to an abrupt end. Subsequent events led to further increases in oil prices near the end of the 1970s.

Increased oil prices contributed to increased operating costs for farmers and for the grain marketing system. However, increased oil revenues contributed to increased demand for grain exports. Newly-rich oil producing countries had ready access to foreign exchange for grain purchases. Increased oil revenues also increased the supply of credit available to less developed countries as international bankers recycled the larger export earnings.

Price Policies in Centrally Planned Economies

The centrally planned economies, most notably the Soviet Union, Eastern Europe, and China, were continuing the policies during the 1970s of maintaining fixed prices at the consumer level for agricultural products. Large consumer subsidies were required in the Soviet Union and Eastern Europe when those countries bought grain on the world market and then sold it to consumers at lower prices. Artificially low consumer prices tend to encourage more consumption than would otherwise occur. Eastern Europe in particular imported large quantities of grain, by historical standards, in order to maintain fixed prices at the consumer level.

Figure 2.4 • Total Grain Imports (former Soviet Union, Eastern Europe)

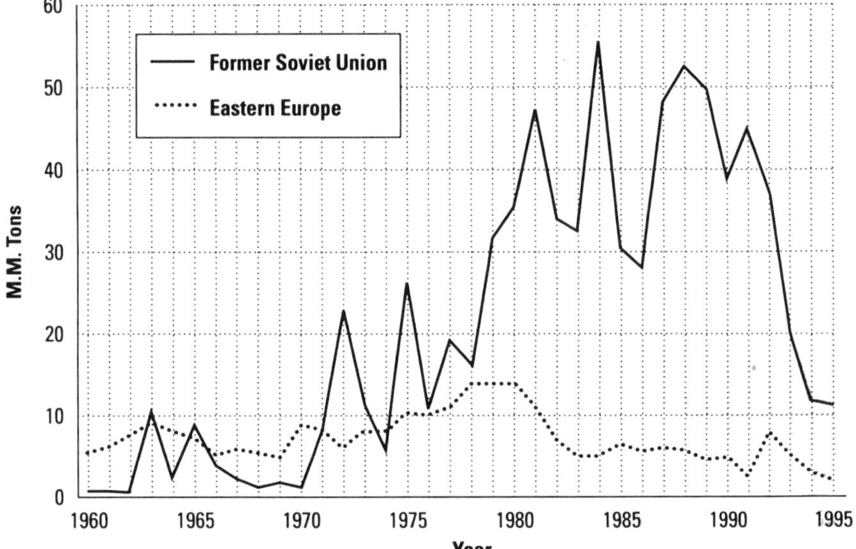

In 1970–72 Eastern Europe imported from 6 to 8 million tons of grain per year (Figure 2.4). Imports rose to a peak of almost 14 million tons during the 1979–81 period. A large part of these imports was purchased on credit which was later defaulted or forgiven.

This same price policy of consumer subsidies was an important factor in the rapidly growing grain import market in the Soviet Union. This policy meant that consumers in centrally planned economies were shielded from increases in world prices for food. Consumers were not required to make the same reductions in food consumption that were required in many less developed countries when world grain production dropped below trend in 1972 and 1974.

U.S. Responses to Growth in Export Demand

Use of Stocks

As noted above, the immediate response to the surge in demand was made by selling grains already in storage. The CCC had sold all its stocks of wheat by the end of 1974. Likewise, stocks of other grains were drawn down. Stocks-to-use ratios for both corn and wheat were reduced to a level last seen during World War II. This reduction in stocks and concern over world grain prices and consumer food prices were the main driving forces that brought about the World Food Conference in Rome in 1974. Concern over stocks also generated a large amount of research on the role of publicly held reserve stocks as a component of food and agricultural policy in the United States.

Return to Full Production

Throughout the 1950s and 1960s and through 1972, from 50–60 million crop acres were held out of production in the United States under various government programs designed to enhance income to grain farmers and hold down the cost of price support programs. Land was idled on both an annual basis and a long-term basis under various programs. When government stocks were depleted, this land was allowed back into production. For the 1974 and 1975 crop years, no land was idled under government programs.

Additional land in production and additional demand for grain led to massive investment on the part of farmers in new, larger machinery, buildings, and storage facilities. After the CCC had disposed of its grain stocks, it divested itself of government-owned storage facilities. The grain marketing system began a storage expansion phase, investing in new larger elevators and acquiring through lease or ownership fleets of railroad hopper cars for transport of grain. Availability of credit, new developments in grain handling and storage technology, and the concept of unit trains for transporting grain to port facilities combined to create an atmosphere of optimism in the grain industry.

Total grain storage capacity increased from 16.6 billion bushels in 1978 to 23.2 billion in 1988. In 1978, 40.1 percent of this capacity was off-farm and 59.9 percent in on-farm facilities. Total capacity increased 40 percent over the 10-year period with off-farm storage at 9.6 billion bushels, accounting for a slightly larger share of total capacity (41.3 percent in 1988) (Figure 2.5). Total storage capacity continued to increase after 1980 when exports reached their peak (Figure 1.1, Chapter 1). Ending stocks of grain reached a peak in 1986 and had dropped sharply by 1988.

Expectations that export growth would continue, accompanied by federal income tax policy incentives and government storage payments, encouraged the growth in storage capacity both on- and off-farm which continued well past the peak in demand for those facilities. This meant many units of storage capacity went unused after the mid-1980s.

Figure 2.5 • U.S. Grain Storage Capacity

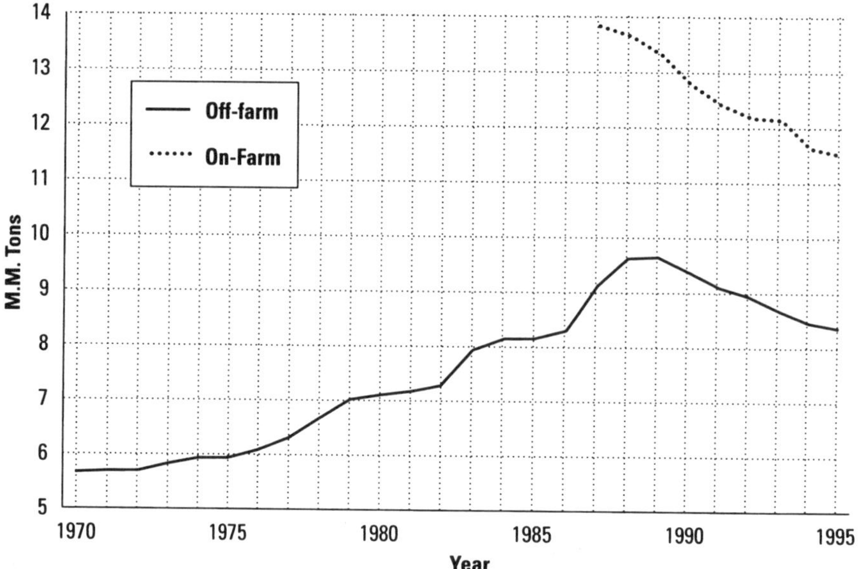

Commodity Policy Changes

Throughout the 1960s, grain prices fluctuated in a narrow range just above the established loan rate. The loan rate put a floor under grain prices, and a combination of large stocks held by CCC along with its release price policies tended to put a ceiling on prices just a few cents above the loan rate. From 1972 to about 1978 the loan rate was well below market prices that had risen as a result of the export boom. Since the CCC did not own stocks during this period, prices were two to three times the loan rate level. But this situation was of short duration. Passage of the 1977 Agricultural Act and expanding grain production in response to higher market prices brought this relationship to an end. Farmers' interest in getting support prices up to cover new, higher costs of production was a factor in getting loan rates raised in the 1977 Act. The Act set minimum target prices for grain and required the Secretary of Agriculture to annually adjust them upward to reflect changes in the cost of production. The increase in loan rates set the stage for renewed acquisition of grain by the CCC through loan forfeiture when market prices failed to exceed loan rates plus carrying costs.

The 1977 Act authorized a farmer-owned grain reserve. Grain was to be accumulated in the reserve when prices were relatively low. Ownership of the grain was to be retained by farmers who would store grain on their farms or in commercial facilities for periods of not less than three or more than five years. Participants would receive a non-recourse loan on the commodity from the CCC. Grain could be put in the reserve at the end of the initial nine-month storage period or sooner if authorized by the Secretary of Agriculture. Grain would be released from the reserve as prices rose above specified levels. Farmers would receive storage payments and the Secretary could waive interest charges on grain in the reserve.

The higher loan rates and the option of continuing to store grain with most of the cost paid by the CCC provided incentives for grain to go into storage and stay there rather than being sold at a lower price in the world market. The government was getting back into the business of public storage of grain.

Exports Decline: Contributing Factors

The decade of U.S. export growth in the 1970s was followed by a sharp and continued decline through 1985 (Figure 2.3). After 1985 U.S. grain exports increased each year through 1988 then again declined by a fairly small amount. U.S. soybean exports reached a peak in 1981 and have shown an irregular downtrend since that date. There are a number of reasons offered for the decline in U.S. exports. U.S. monetary policy to control inflation, worldwide recession, the 1980 U.S. grain embargo, U.S. policy response as continued in the 1980 Act, the collapse of Eastern European demand and growing production in the European Community are all cited as developments which contributed to the decline in U.S. grain exports. Each development is briefly discussed below.

U.S. Monetary Policy to Control Inflation

Compared to previous decades, the United States experienced unprecedented levels of inflation in the 1970s. The country was used to price level increases of 2–3 percent per year up to the early 1970s. After 1972 annual inflation rates ranged as high as 13.5 percent per year. During most of this time the Federal Reserve used interest rate policy as its primary tool for both stimulating the economy and controlling the rate of inflation. Failure of this policy to control inflation led to a focus on growth in the quantity of money as the primary tool for controlling infla-tion. This policy was initiated by the Federal Reserve System in October 1979. Interest rates were allowed to seek their own level. Subsequently, in 1980 the prime interest rate rose from about 12 percent to 20 percent. A shock of this magnitude soon brought inflation rates down as economic activity slowed. Inflation rates have continued down and have averaged from three to four percent per year.

During this period, the value of the dollar rose sharply in response to the higher interest rates, gaining about 50 percent in value relative to a basket of major currencies. A rising value of the dollar has an effect on exports which is opposite to that discussed above under monetary ex-change rates. With a stronger dollar, U.S. grain increases in price when measured in foreign currency. To the extent that these price increases are transmitted to the buying country, they tend to decrease U.S. ex-ports.

Worldwide Recession

The effects of tight monetary policy in the United States in the early 1980s were soon transmitted around the world. As other economies ei-ther stagnated or grew at much slower rates than in the 1970s, demand for grain imports declined. West Germany and Pacific Basin countries continued to grow but most other countries experienced slow rates of growth in the early 1980s.

U.S. Grain Embargo

The trade embargo against the Soviet Union which was imposed in early January 1980 is considered by some observers to have had a large nega-tive impact on U.S. exports. The embargo was imposed by the United States in response to Soviet entry into the civil war in Afghanistan on the side of the Communists. Under the embargo, the Soviet Union was al-lowed to receive grain already in transit. This amounted to about 8 mil-lion tons called for in the United States–Soviet Union long-term agree-ment. All other sales were canceled. An action involving this volume of grain severely disrupted the market in the short run.

The United States initiated a series of actions in early 1980 to reduce the impact of the embargo on farmers and on exporters who had already purchased grain to fill the canceled contracts. Many farmers to this day believe they suffered uncompensated damage as a result of the embargo. A major study authorized by Congress does not support popular opinion. It concluded that actions by the U.S. government to compensate farmers and exporters minimized their potential damage. An unresolved issue was the long-run effect of the embargo on future trade relations and on the reputation of the United States as a reliable supplier.

An unanticipated event minimized the cost of the embargo to the government and to farmers. U.S. corn production and spring wheat production were down in 1980 as a result of widespread drought in the Midwest and Great Plains. Grain prices rose sharply as a result of reduced supplies. Grain not delivered as a result of the embargo was available for sale at prices above those in place at the time of the embargo.

U.S. Commodity Policy During the Early 1980s

The Agriculture and Food Act of 1981 continued the system of price and income support contained in the 1977 Act. Loan rates continued to be the key component of the income support system. Loan rates were set at a fairly high level and target prices for the 1982–1985 crops were established on a rising scale anticipating rising costs for the production of grain.

The ink for the 1981 Act was barely dry when the economic environment for agriculture changed dramatically. The annual U.S. inflation rate, measured by changes in the Consumer Price Index, reached a peak of 13.5 percent in 1980 then declined to 1.9 percent by 1986 as a result of the actions taken by the Federal Reserve Board. The prime interest rate charged by banks increased to 18.9 percent in 1981 in contrast to the 6.8 to 12.7 percent rates experienced during the 1970s. As noted, the exchange value of the dollar relative to major currencies increased by over 50 percent in a period of five years.

Price support levels contained in the 1981 Act stimulated high levels of program participation, large grain takeovers as non-recourse loans were defaulted and growing stocks in the hands of the CCC. The United States. was in effect pricing itself out of the world market.

A large U.S. corn crop in 1982 and declining exports led to more drastic measures for controlling production in 1983. The program was so attractive to producers that they signed up in unexpectedly large numbers, idling 70 million base acres from grain production. Participants received a part of their payments in Payment-in-Kind (PIK) certificates which they could exchange for grain owned by the CCC or under loan with the CCC. This provided a way to release grain from government storage while maintaining the rigid resale rules established by Congress.

U.S. corn production in 1983 was reduced to just over 4 billion bushels compared to an 8 billion bushel crop in 1982. A severe drought was responsible for about half the reduction, with idled acres accounting for the other half. The 1983 PIK program inflicted losses on the grain marketing system as it adjusted to the much smaller volume of grain from the 1983 crop.

Collapse of Eastern European Demand

After reaching a peak of about 14 million tons of grain imports in 1980–81, Eastern European imports collapsed and by 1984 were at about the 5 million ton level (Figure 2.4). They have fluctuated in a range from 2.5 to 7 million tons since. Economic stagnation in Eastern Europe, loss of access to credit because of repayment history, and activities leading up to eventual overthrow of the Communist system all contributed to the decline in imports.

Growth of Production in the European Community (EC)

The original six members of the European Community implemented their Common Agricultural Policy (CAP) in 1967. At the time the CAP was implemented, the EC countries were large net importers of grain and oilseeds (Figure 2.6). The CAP featured high support prices relative to world trading prices for grain, a variable levy system to keep imported grain prices in line with domestic prices, and a system for buying producers' grain in order to maintain established target prices. The system of price supports and import restrictions provided a powerful stimulus to EC grain production. The EC expanded to include 12 members in 1986. At that time, EC population was very similar in size to U.S. population. Likewise, gross domestic product for the two economic entities has become quite similar in size. The EC-12 went from being net importers of 20 to 30 million tons of grain per year to zero net imports in 1983. Since that date, they have become larger net grain exporters, with exports in the 20–25 million ton range (Figure 2.6). These developments represent a swing in net imports of over 45 million tons in a period of 30 years. It represents a lost market for U.S. grain exports and increased competition for U.S. grains in third country markets. The EC has to use large export subsidies to move its export surplus into the world market.

Figure 2.6 • EC-12, Net Grain Trade

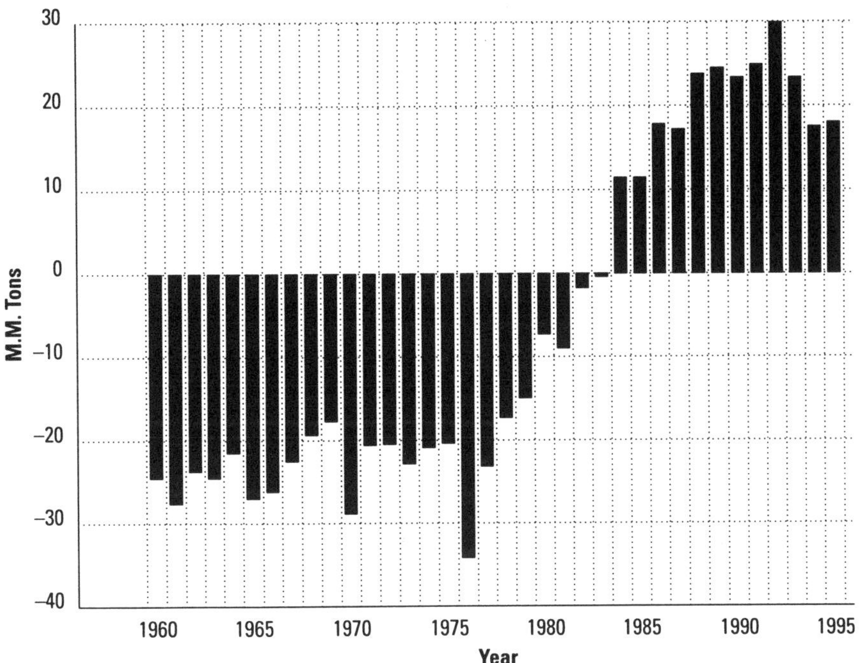

United States' Response to the Export Decline

Reorientation of Commodity Policy

The Food Security Act of 1985 continued the loan rate–target price approach for commodities. But it was designed to be much more friendly to grain exports than the 1981 Act. Target prices were established at a relatively high level to provide income protection to participants. Loan rates were lowered by law with authority given to the Secretary to lower them further if necessary to make U.S. grain competitive in world markets. The Secretary was authorized to use generic payment-in-kind (PIK) certificates for 17 different program payments authorized by the Act. Participants could use the certificates which were denominated in units of grain to redeem grain from storage (their own under loan or from the CCC) or sell them for cash. A subsequent holder could redeem them for CCC grain.

This system encouraged increased participation by farmers in voluntary farm programs as it in effect removed the loan rate floor under grain prices. The system allowed market prices to become much less restricted on the downside, provided income protection for participating farmers, and allowed grain to flow out of government storage. Non-program participants felt the effects of lower market prices for their grain but were not entitled to the deficiency payments which were a part of the program.

The Export Enhancement Program (EEP) was authorized and funded as a means of subsidizing grain into the world market. Exporters were issued PIK certificates which they could use to acquire enough grain to lower the average cost of grain they had purchased on the market. Initially, the EEP was to be used to subsidize grain into markets where the EC was actively competing with their use of subsidies. The program came to be used in many other situations and may have been used as a bargaining chip in ongoing international trade negotiations under the General Agreement on Tariffs and Trade (GATT). The 1990 Act provided for expansion of the EEP if an agreement was not reached by a specific date (a date which has passed).

Longer-Run Prospects

After seven years of difficult trade negotiations under auspices of the General Agreement on Tariffs and Trade (GATT), agreement was reached in late 1993. The agreement was accepted by the major participants in 1994 and began to be implemented. Signing of the agreement and several related developments suggest that price levels will become less an issue in trade negotiations and the amount of government support for agriculture will likely decline, moving the system more toward a market system.

With the Food, Agriculture, Conservation and Trade Act of 1990 the United States, while maintaining price support levels during the first years of the program, lowered its price support levels for grain for the out years and moved more toward direct payments to participants, a policy that was visualized as interfering less with trade than a high price support system. The 1990 Act provided for reduction in income payments for 1994 and 1995 crops. The Federal Agricultural Improvement and Reform Act (FAIR) of 1996 provided income support to participants for seven years starting in 1996. That Act, while continuing provisions for non-recourse loans for grain at relatively low levels, eliminated almost all restrictions on production. Participants were to receive an annual payment based on historical acreage bases and yields. These payments are to decline from the 1996 level to 2002 by about 29 percent. The so-called Freedom to Farm Act gave farmers the freedom to respond to market signals but essentially eliminated the safety net under grain prices. Authority for use of export subsidies was retained at a scaled-down level.

The European Community (now the European Union or EU) agreed to reduce its support levels by a significant amount over the next few years. The EU also agreed to reduce the quantity of grain it will subsidize into world markets. The agreement calls for a step-wise reduction over a period of years in the quantity of grain the EU subsidizes into world markets. Since production capacity has grown so dramatically since 1960 and investments have been made in production facilities, it seems unlikely, however, that EU producers will cut their production sufficiently to again become net importers of grain in the foreseeable future. Experience in the U.S. with asset fixity in agriculture suggests it is much more difficult to reduce production than it is to increase it.

Dramatic economic and political changes occurring in the former Soviet Union and in Eastern Europe raised questions about future export markets there. It is not beyond a reasonable doubt that they may become competitors for world grain markets given their land resource base and history of grain production prior to adopting the Communist system. Observers who have "been there" focus on the needed adjustments, the poor state of infrastructure, and the attitude of the people toward a market-oriented system. They believe the adjustments will be slow, painful, and not without danger of reverting to the former system.

World grain trade rebounded after hitting a low point in 1985. World trade in grain reached a new peak in 1992 and then declined again. Although total U.S. grain trade also increased after 1985, it has not yet returned to its peak level of 1980. While U.S. grain trade is still below its peak of 1980, the value of total U.S. agricultural trade reached a new peak in 1996 at almost $60 billion. The composition of U.S. agricultural trade has changed significantly since 1985. Bulk grains have become a less significant part of that trade while meats, poultry, horticultural products and processed foods have become more important. This growth has been a result of rising consumer incomes world-wide, especially in Pacific Rim countries and the competitive position of U.S. food processors. The changing composition of U.S. agricultural exports may be a factor in explaining why traditional grain companies are becoming more integrated into livestock production and the food marketing system.

In summary, several observers expect modest growth in the demand for U.S. grains and oilseeds over the next decade. Howevr, recent growth in export demand for the value added products—meat, poultry, horticultural products, and processed food prodcts—will likely slow or even decline as a result of recession or slower growth in Pacific Rim countries. After a time, these economies will likely recover and U.S. export demand will again expand. Although export demand will growth, it is unlikely that the same set of forces will evolve in the same sequence to produce an export boom like the one experienced in the 1970s. However, major weather changes could generate short-run unexpected demand for U.S. grains and oilseeds.

References

Cotterill, Ronald, *Cartels and Embargoes as Instruments of American Foreign Policy*, Agricultural Economics Report No. 373, Michigan State University, East Lansing, April 1980.

Gudmunds, Karl and Alan Webb. Time Series Annual Data, PS & D, 1992, USDA, ERS.

Kitchen, John and Ralph Monaco. "Effects of Macroeconomic Policies on Agriculture," *Agricultural-Food Policy Review, U.S. Agricultural Policies in a Changing World*, USDA, ERS, Ag. Ec. Rept. No. 620, Nov. 1989, pp. 27-36.

Oehrtman, Robert L. and L.D. Schnake, "Marketing Channels and Storage," Chapter 3 in *Grain Marketing*, Gail L. Cramer and Eric J. Wailes (eds.), Second edition, Westview Press, Boulder, Colorado, 1993.

U.S. Dept. of Agriculture, Economic Research Service, "Embargoes, Surplus Disposal, and U.S. Agriculture," Staff Report No. AGES 860910, November 1986.

3

Futures, Options, and Cash Markets

William W. Wilson, David W. Bullock,
and Bruce Dahl

Introduction

There have been dramatic changes in the market system for price discovery and risk management in the United States grains sector during the past 20 years. This is reflected in the rapid escalation in futures trading and the introduction of options trading for agricultural commodities. The purpose of this chapter is to describe these changes. Sections 2 and 3 describe the growth in futures trading and the evolution of options trading. Section 4 describes the multitude of different contracts that have emerged in grain marketing transactions. Section 5 provides a discussion of important issues that are emerging at the U.S. grain exchanges.

Increases in Futures Trading

The volume of trade occurring in many of the futures contracts for grains, livestock, and other futures contracts (metals, financial instruments, and market indexes) from 1980 to 1997 has increased dramatically. However, changes have varied by contract and sector. This section describes those trends. The data are summarized in Figures 3.1-3.6 and Table 3.1.

In the grains sector, futures trading is dominated by transactions for Chicago Board of Trade (CBOT) corn and soybean contracts with soybeans traded more in the 1980s and corn more in the 1990s. The average daily volumes for these two contracts have varied significantly from year to year and are 2 to 3 times the volume of trade for the next most active grain futures contract (CBOT Wheat). The volume of trade for selected futures contracts in the grains sector largely showed declines from 1980 to 1990 that ranged from a low of –6 percent for CBOT corn to a high of –48 percent for CBOT Wheat. The only exception to this trend was for the Minneapolis wheat futures contract which increased average daily volume by 40 percent from 1980 to 1990.

In the 1990s, the volume of futures trading for the grains sector has increased dramatically, reversing the trend toward lower volumes experienced in the 1980s. Average daily contract volumes increased for all the major futures contracts in the grains sector with much of the increase in trade volumes occurring since 1994. The increase in volumes from 1990 to 1997 ranged from a low of 41 percent for CBOT soybean futures to high of 115 percent for Minneapolis wheat.

Trade volumes for futures contracts in the livestock sector have largely declined or remained stable. Trade volume for CME live cattle declined 36 percent from 1980 to 1990, but increased 4 percent from 1990 to 1997. The trade volume for CME live hogs increased 4 percent from 1980 to 1990, but has declined 6 percent in the 1990s.

Trade in many of the financial, metal, and index futures has been explosive from 1980 to 1997. In particular, trade in CME Eurodollars has grown to over 400,000 contracts traded per day in 1997. This is greater than the volume of futures contracts traded for many of the grains and livestock futures contracts combined. Trading for CME S&P 500 futures have increased from limited volume in 1980 to near 100,000 contracts per day in 1990 and have since increased to 150,000 contracts per day by 1997. Similarly, trading for NYMEX crude oil has increased from nominal trading volumes in the early 1980s to 100,000 contracts per day in the 1990s. Trading in NYSE natural gas has also increased dramatically in the 1990s while COMEX silver has shown a steady increase in the 1980s and 1990s. Meanwhile other contracts like the CME 3 Month Treasury Bill and KCBOT Value Line have had marked declines in trading volumes in the 1980s and 1990s.

Table 3.1 • Percentage Change in Average Daily Volumes and Open Interest Between 1980-90 and 1990-97 for Selected Futures Contracts.

Exchange/Contract	1980-90		1990-97	
	Volume	Open Interest	Volume	Open Interest
	%			
CBOT Soybeans	–13.7	–21.7	41.7	46.5
CBOT Corn	–5.8	3.2	48.1	48.0
CBOT SRW Wheat	–48.0	–7.9	77.0	56.5
KCBOT HRW Wheat	–13.2	–6.7	70.4	74.1
MGE HRS Wheat	40.7	32.0	115.7	91.2
CME Live Cattle	–36.9	41.4	3.5	21.3
CME Live Hogs	3.5	17.9	–5.7	7.5
NYMEX Crude Oil	NA	NA	4.5	49.0
NYMEX Natural Gas	NA	NA	6498.6	2392.5
COMEX Silver	269.0	125.7	26.7	–1.8
CME Eurodollars	NA	NA	205.2	286.7
CME 3-Month Treasury Bills	–44.5	29.2	–89.0	–73.6
CME S&P 500	NA	NA	57.8	49.8
KCBOT Value Line	NA	NA	-60.0	–76.1

Options on Futures and Their Role in Risk Management

One of the most important innovations that has been introduced into the grain marketing system has been options. These are important because through their use a multitude of different risk management strategies can be pursued and/or numerous types of contracting mechanisms can be developed. This section describes the evolution of option trading and discusses some of the strategies that market participants can pursue.[1]

History of Options on Grain Futures

Agricultural options provide a risk management alternative for most participants in the agricultural marketing chain. Grain options began trad-

Figure 3.1 • Average Daily Volume for Selected Grain Futures, by Year

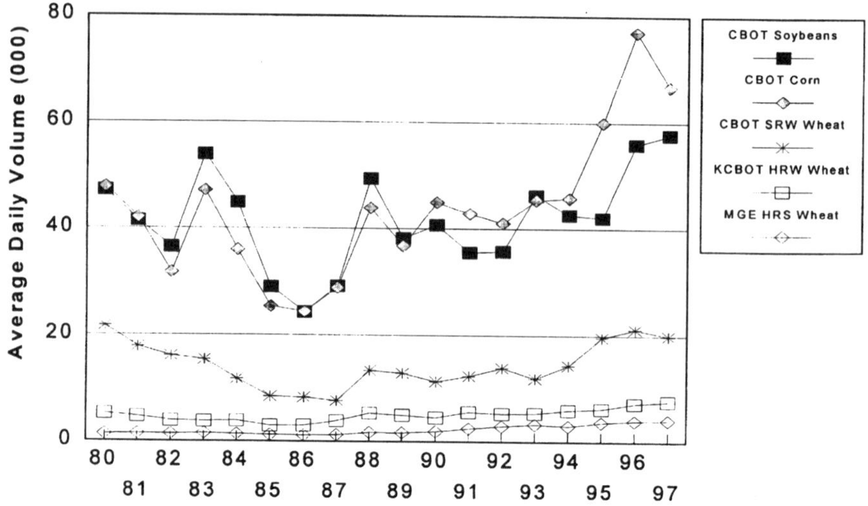

ing on the Chicago Board of Trade as early as the Civil War. Shortly after their introduction, however, the Board of Trade tried to halt the trade in such options because of perceived abuses. Illinois banned option trading both on and off exchanges, but trading continued despite this statutory ban. In 1887, the Board of Trade restricted option trading among its members. Agricultural options were periodically traded and banned between 1887 and 1936, at which time Congress completely banned the trading of agricultural options by enacting the Commodity Exchange Act of 1936.

A number of factors precipitated the 1936 ban, including the following: Options were not being used for traditional risk-shifting purposes, small traders "lured" into options markets to speculate usually lost money, large traders could use options to cause artificial price movements; terms and conditions of options were not standardized, and congestion near the close of the market could occur because many options were good for less than a day.[2] A specific case occurred in the early 1930s when options were blamed for excessive price movements in wheat, which led to the collapse of the Chicago Board of Trade's wheat market in 1933. Problems continued to occur in the late 1960s and early 1970s when gold options were being sold fraudulently.

The problems seemed to stem from the lack of one specific governing body. After the gold incident, the Commodity Futures Trading Com-

Figure 3.2 • Average Daily Volume for Selected Livestock Futures, by Year

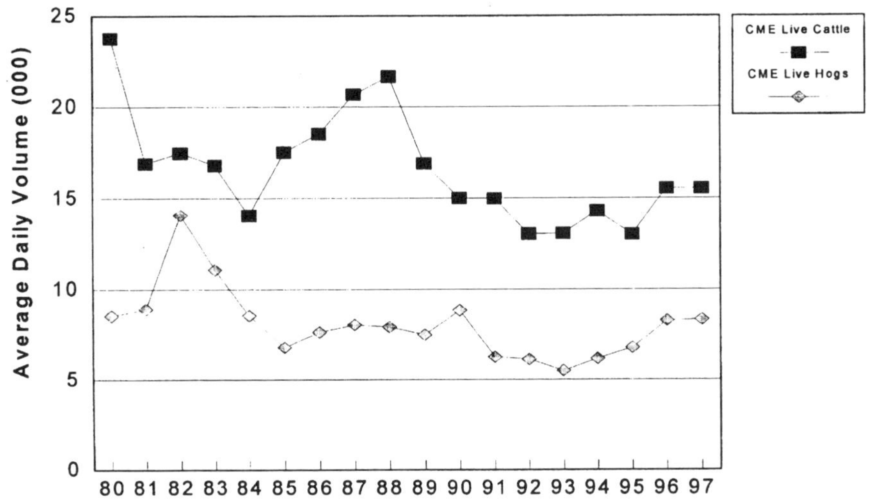

mission (CFTC) was given full control of all commodity futures and options trading. The CFTC worked to develop regulations for exchange traded options and, in September 1981, approved a three-year pilot program, including options trading in gold, treasury bonds, and sugar. Following this, the Futures Trading Act of 1982 lifted the 1936 ban on options on agricultural futures. The result was the development of an experimental three-year program of organized trading in selected agricultural options, which was introduced in October 1984. That development made options an integral component of the agricultural marketing complex and since has been fully adopted into the market system.

Use of Options by Producers and Merchants

Traditionally, elevators, domestic merchants, processors, and exporters have used the futures market extensively to hedge cash positions. Producers hold similar cash positions, and the same principles apply. They can use the options market as a substitute and/or supplement to futures market for hedging. Each of these participants performs unique functions in the grain marketing chain. However, all are similar in the sense that they can have similar cash positions. This similarity allows for generalizations across producers and merchants about the effects of alternative option transactions used with alternative underlying cash positions.

Hedging Long and Short Cash Positions. For a merchant with a long cash position, the greatest risk is that associated with the price upon sale of the grain. A producer has similar risk when planting or storage decisions are made. The only difference between a producer's and a merchant's cash position is that before harvest, the producer also has an element of risk associated with production (i.e., weather-related yield risk). Both participants want to protect the value of their cash market position. When the futures market is used, a future price is "locked in," thereby minimizing risk. However, they would be unable to benefit if futures prices increase after the futures position is taken. In this situation, an option contract would be desirable. Merchants or producers with long cash positions can either buy puts or sell calls.

Long Cash/Long Puts. A merchant or producer with a long cash position, who wants protection against a bearish price move, could lock in a floor price and also benefit from a potential price rise by purchasing put options. A long put position gives buyers the right, but not the obligation, to assume a short futures position at the exercise price. However, instead of either exercising the option or letting it expire, the buyer also may offset the position by selling the puts. The price movements of the underlying futures and corresponding options affect the buyer's deci-

Figure 3.3 • Average Daily Volume for Selected Commodity/Index Futures, by Year

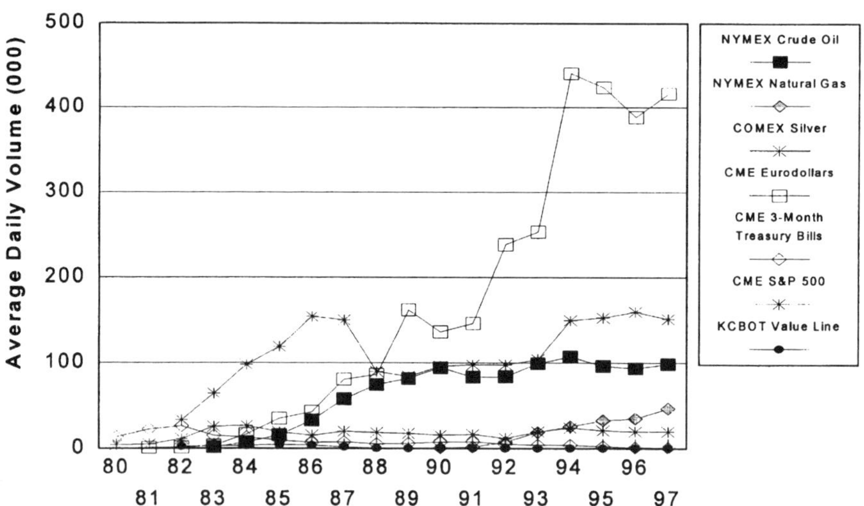

sion. If the underlying futures price increases significantly over the life of the option, the buyer could either let it expire and realize the greater cash price or sell the puts, which would have fallen in value with increased futures.

If the opposite occurred, that is, the underlying futures price dropped, the buyer could again pick one of two possibilities: exercise the option and assume a short futures position at the strike price or offset the option, receive a premium, and be out of the futures market completely. In either case, the risk is limited to the premium paid while possible returns are unlimited.

Short Cash/Long Calls. A merchandiser with a short cash position could establish a ceiling price and take advantage of a fall in prices. A common scenario is for a merchant or processor to sell grain for some future month, but not have the physical grain on hand. Purchasing call options would establish a ceiling price for the eventual cash purchase. Merchants can either buy calls or sell puts against a short cash position.

A merchant or producer with a short cash position who wants to protect against a bullish price move and benefit from a potential price decline could lock in a ceiling price by purchasing call options. Buying calls is preferable to traditional hedging if futures prices decrease be-

Figure 3.4 • Average Daily Open Interest for Selected Livestock Futures, by Year

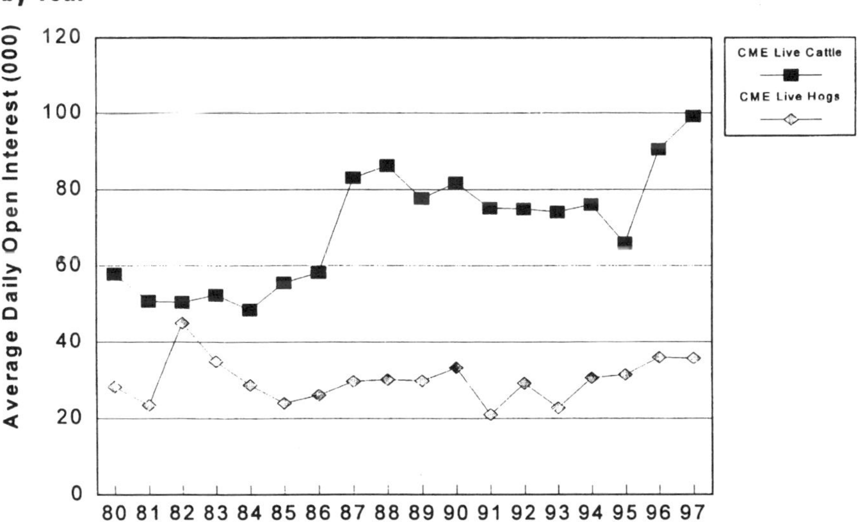

Figure 3.5 • Average Daily Open Interest for Selected Grain Futures, by Year

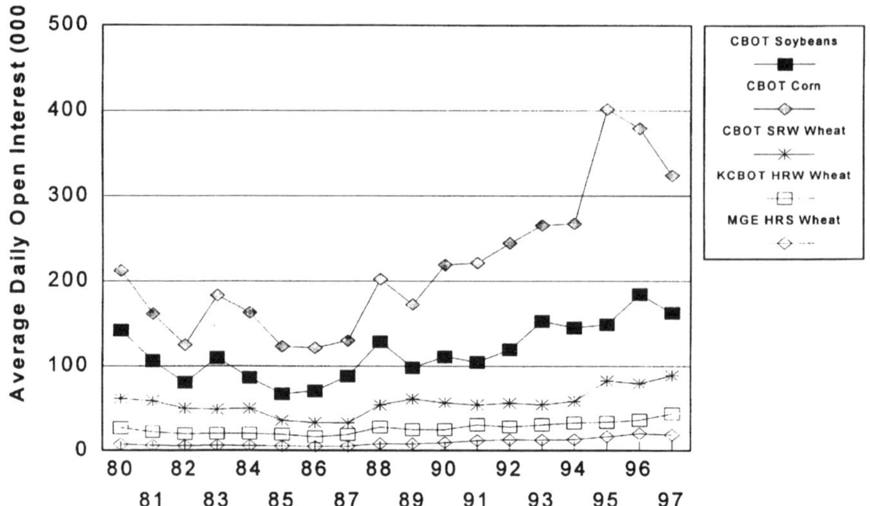

cause losses in the futures do not offset the gain in the cash price. If futures prices increase, buying futures yields a greater return by the cost of the option premium.

Long Cash/Short Calls. Anotherr alternative for a merchant/producer with a long cash position is to sell call options. A short call position allows the seller to augment current income by the amount of the premium and is particularly attractive if prices are exprcted to remain relatively stable. In this case the return is limited to the premium.

A merchant or producer with a long cash position has two basic option alternatives: to buy puts for locking in a floor price or to sell calls to supplement current income. An important difference between the two choices is the risk level involved. Buying puts is preferable to traditional hedging if fuyures prices increase because losses in the futures do not offset the gain in cash prices. A long put position limits risk and provides unlimited profit potential, while a short call position limits the return to the premium.

Short Cash/Short Puts. Buying call options is not the only strategy for a merchant or producer with a short cash position. Put options also could be sold. The best outcome would occur if the market remains relatively stable so that the options would not be exercised, but yet would retain some intrinsic value. Selling puts would produce a limited return equal

Figure 3.6 • Average Daily Open Interest for Selected Commodity/Index Futures, by Year

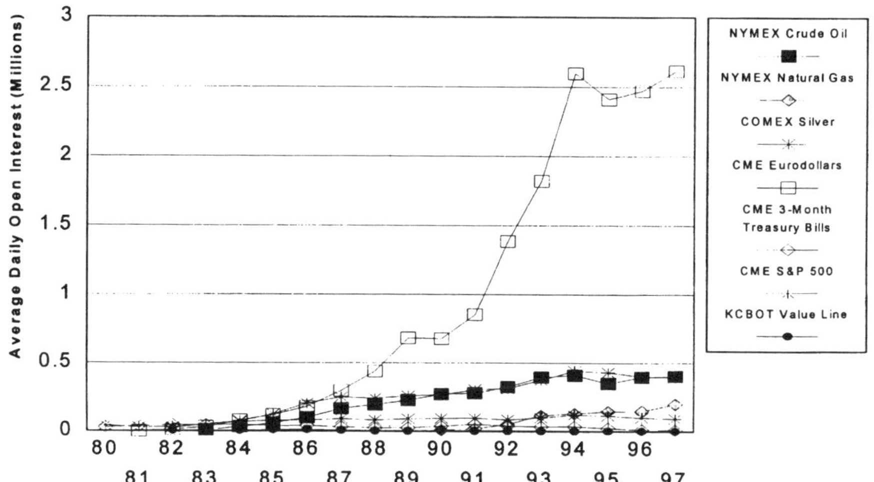

to the premium. In addition, the merchant is required to have margin money available.

Overall, a short put position has virtually the same fundamental characteristics as a short call position. Sellers can choose only between liquidating their position before the puts are exercised or waiting for the buyer's actions.

Forward Contracts With Floor and Ceiling Prices. One of the very important virtues of options is that they facilitate development of alternative contractual relationships. Most prominent among these has been the development of forward contracts with floors or ceiling prices. Typically, forward contracts are made by elevators to producers, by exporters to importers, and by merchants to processors. In the past, these types of contracts have been implemented via the futures market. The result has been a forward contract with a fixed price. The disadvantage of this is that a producer would not benefit from a rise in prices and an importer would be unable to take advantage of a fall in prices. Agricultural options stand to change this scenario by making forward contracts more attractive to producers and end users.

Merchants can offer a forward contract with a floor or ceiling price by incorporating the use of options. For instance, an elevator can lock in a floor price for its own grain sale by purchasing put options and then pass this floor price on to producers through a forward contract. The cost of the put options, or the premium, would be included in calculating the producer's floor price. Likewise, a merchant could lock in a ceiling price on purchases by buying call options, which could then be passed on to end users via a forward contract. Once again, the ceiling price would reflect the exporter's cost of buying the calls.

A number of advantages to both merchants and their customers can be realized by incorporating options into forward contracts. The biggest advantage to producers and end users is that they could lock in a floor or ceiling price and could benefit from a favorable cash price move; with traditional forward contracts, they would not. Furthermore, the new forward contracts would permit individuals to take advantage of options indirectly through merchants without being directly involved in the option market or needing to understand the mechanics themselves. Consequently, forward contracts based on options offer another merchandising alternative, which has potential applications throughout the grain exchange system. Forward purchase contracts have typically been made at fixed prices. One of the most common transactions with this type of contract is between an elevator and a producer.

Other Applications: With the advent of options trading, additional applications exist for risk management in grain marketing, notwithstanding the increase in alternatives for speculation. Three additional grain-marketing applications using options are discussed here.

Producers store grain after harvest for a number of reasons, one of which is the potential appreciation in postharvest prices. Storing grain for this purpose is essentially speculative, but is a widespread practice. A problem, particularly during the early 1930s, was that grain storage often resulted in net losses due to the high opportunity cost of capital. However, selling at harvest precluded benefits of postharvest price appreciation. An alternative, using options, would be to purchase calls concurrent with the cash sale of grain.

If price appreciated, so would the call, while storage and interest costs would have been saved. A cash price decrease would result in a lower call value, but the loss would be limited to the initial premium. Analyzing this alternative requires a careful comparison of storage and interest cost to the cost of the premium. Savings in storage and interest may offset the cost of the call premium.

Another alternative that was attractive during the late 1980s was to use options to protect against deficiency payment reductions. During this time, grain prices were escalating, causing the potential for deficiency payments to decrease. By purchasing call options, producers could increase their revenue so long as futures increased. This increase then could offset a lower deficiency payment associated with higher cash prices.

Another use of options is as an alternative to hedge against quantity risk, which imposes problems for traditional hedging for various participants involved in grain marketing. Merchants are exposed to quantity risk after having tendered an offer to buy or sell grain. A long option position can be used to hedge against this risk by limiting the possible loss to the premium cost. If the futures market was used, the potential futures loss of an adverse price move is unlimited. In this sense, options are more appropriate than futures for relieving risk when quantity risk is present.

A merchant who has tendered an offer to sell grain, but will not know for several days whether the offer will be accepted, can buy call options to hedge this position. If the offer is accepted, the exporter would have short cash/long calls. If the offer was rejected, the results again would depend on the futures price. The largest return in this scenario would be realized with an increase in the futures price. An elevator also may post a price to buy grain with the offer remaining valid for a specified number of days (e.g., overnight or over a weekend). In addition, the estimated purchase may only be partially fulfilled. Put options can be bought as a hedge against this type of transaction.

The maximum loss associated with a long put position is the premium cost. If the offer was accepted, the elevator would have a long cash/long put position. If the offer was rejected or partially fulfilled, the largest return would have occurred with a decrease in the futures price. The premium would not have been recovered if the futures price had increased. However, the loss would not be any greater than the premium. Thus, a long option strategy appears to be more advantageous than a futures hedge when a merchant has tendered an offer for either a purchase or a sale.

Producers likewise are exposed to quantity (yield) risk in preharvest hedging, and, as such, their alternatives are similar to the elevator example (i.e., long cash/long puts). In general, when quantity risk exists for producers, using options may be more appropriate than futures. Using options to reduce risk is generally cheaper than using futures when quantity risk exists. A quantitative analysis of this is fairly extensive and is not presented here.

Hybrid Cash Grain Contracts

The succession of market-oriented farm bills, starting with the 1985 bill and ending with the recent Freedom to Farm Act of 1996, has drastically reduced the pricing safety net available to farmers and has effectively exposed them to the market. However, with this reduction in the farm safety net has come the increased freedom to plant whatever crop the farmer desires. Farmers now have to look to the marketplace to find tools to manage their pricing risk. The country elevator is in an ideal position to offer market pricing tools to farmers and use this array of tools as a way to source grain away from their competitors. Those elevators that are most innovative in their offering of contracts to the farmers stand a greater chance of receiving most of their business.

Traditionally, traders have offered three types of contracts: fixed price, basis, and deferred price. The accelerated emphasis upon product innovation in the past ten years has led elevators to begin offering what are called "hybrid contracts." In addition, proposed policy changes by the Commodity Futures Trading Commission (CFTC) have led to the possibility of reintroducing previously banned "agricultural trade options" into the marketplace.

Hybrid Cash Contracts: Definition and Structure[3]

In the grain merchandising trade, the term hybrid contracts is used to describe various forms of cash contracts between grain elevators and producers that incorporate the value of exchange-traded futures and/or options into their pricing mechanisms. Among the most popular hybrid contracts are the hedged-to-arrive (HTA), minimum price, maximum price, and min/max contracts.

Hedged-to-arrive (HTA) Contracts. A HTA contract, in its most basic form, calls for future delivery of cash grain with the futures portion of the cash price fixed today and the basis portion to be determined at the time of delivery. This is similar to a short futures hedge, where the producer would sell futures through a licensed broker to lock in the futures portion of his cash price risk. The producer would then either offset his futures position or deliver the grain on the contract at the time of delivery, with the basis determining the final sale price.

Under the HTA, the elevator will sell futures at the same time that the contract is written to offset its futures price risk exposure under the contract. In essence, the producer is short hedging his crop indirectly through the local elevator rather than going through a licensed broker.

However, there are some major differences between HTAs and the basic short futures hedge. Under the short futures hedge, the producer is directly responsible for financing and maintaining the margin requirements for the short futures position. Under the HTA, the elevator is responsible for financing and maintaining the margin requirements on behalf of the producer. Under the short futures hedge, the producer directly receives regular feedback on the short futures position through account statements provided by her broker. Under the HTA, the elevator receives the account statements, and feedback to the producer is indirect at best.

The popularity of the HTA from the producer's perspective is that the HTA removes the burden of having to arrange for initial financing and a line of credit for futures margin requirements. In addition, the elevator may have an advantage in acquiring margin financing because of its ability to aggregate positions and realize economy of size efficiencies that are present in using futures contracts. Also, from a producer's perspective, the HTA allows deliveries of odd lot sizes rather than the standardized 5,000 bushels required per grain futures contract. Again, the elevator can aggregate positions to better match the standardized futures contract size.

From the elevator's perspective, a HTA is a tool to source grain in an orderly and timely manner that otherwise might not be available if the producer hedges directly with the futures market. The HTA also offers the elevator all of the advantages of a fixed price forward contract without the need to manage basis risk, which remains with the producer.

In addition to the standard HTA requiring delivery at a certain time, there are other variations of the HTA contract. One such variant allows the producer to "roll forward" the futures month and the subsequent delivery period. This is similar to the time-honored practice of "rolling forward" a futures hedge. However, this increased flexibility to the producer also comes at the cost of an added risk tied to the futures spreads. This risk becomes most acute when the contract allows rolling forward into the next crop year, therefore exposing the producer to the old crop-new crop spread risk that is historically more volatile than the within-crop year spreads.

Another variant of the HTA allows the producer to price the grain in a futures month other than the intended delivery period. This adds a level of speculative spread risk that the producer must assume, and it is proportional to the time spread between the intended shipment period and the cash delivery period. Again, choosing a futures month in a crop year other than that corresponding to the delivery period will greatly enhance the level of risk exposure to the producer.

Another variant of the HTA allows the producer to price/unprice his futures positions at any time. When the producer chooses to "unprice" his position, he exposes himself to the full cash price risk over the unpriced period. Allowing this feature may undermine the legal "forward cash contract" status of the HTA and may possibly lead to the contract's being considered by the CFTC as an illegal "off-exchange" futures contract.

A final variant of the HTA allows the producer to "walk away" from his delivery obligation and price out of his contract. This feature will most likely cause the contract to be viewed as an illegal "off-exchange" futures contract by the CFTC.

Minimum, Maximum, and Min/Max Pricing Contracts. A minimum pricing contract either establishes a minimum cash price or a minimum "HTA" (futures only) price for delivery of the cash grain. Under a minimum cash price contract, the elevator manages both the futures price and basis risk exposures for the producer. Under a minimum HTA contract, the elevator manages just the futures price risk component, and the producer manages the basis risk. The elevator usually purchases an offsetting put option to cover its futures price risk exposure. The producer pays the elevator for the cost (premium) of the put option plus a service fee (which includes handling and basis risk management). The basis level and/or the HTA (futures price) may be adjusted by the producer which will impact the option cost and service fee. The elevator may also offer a maximum price contract to its downstream buyers by using call options to place a cap on its selling price.

A maximum price contract to producers is a contract calling for delivery of the grain at a future date at a maximum cash price or HTA (futures price). The elevator writes call options to cover the futures portion of its exposure. For entering this contract, the elevator forwards a rebate to the producer that equals the premium received on the call option net the service fee.

A min/max price contract calls for the producer to deliver the grain with both a minimum and maximum cash price or HTA (futures price) established. This is basically the cash market version of the popular option "fence" or "window" hedge that uses purchased puts to establish a price floor and written calls to establish a price ceiling with the premium received from the call used to offset all or part of the premium cost of the put. In a min/max contract, the elevator is placing a fence position on behalf of the producer, who pays the elevator for the net option cost (put premium minus call premium) and the servicing fee. The minimum and maximum price levels can be adjusted to achieve a desired net premium cost.

The advantages of the minimum, maximum, and min/max contracts to the producer and the elevator are basically the same as for the HTA contracts. The elevator can achieve cost and position matching efficiencies by aggregation, and some of these efficiencies can be passed on to the producer. The elevator also assures itself of a reliable source of grain deliveries.

Agricultural Trade Options: Definition and Structure

In their most general sense, *trade options* are cash commodity contracts between a buyer and a seller who both have valid commercial interests in the underlying commodity. Physical delivery of the commodity is an option rather than an obligation. Under the provisions of the Commodity Exchange Act, the offering of trade options in the agricultural commodities enumerated under the Act is prohibited. However, in late 1997, the Commodity Futures Trading Commission (CFTC), which has administrative authority over the Act, forwarded a proposal for a three-year pilot program to study the lifting of the prohibition of trade options in the enumerated commodities (17 CFR Parts 3, 32, and 33).

Under the proposed pilot program, agricultural trade options which, if exercised, result in delivery of the commodity and which may not be resold, repurchased, or otherwise canceled other than through the exercise or natural expiration of the contract, will be allowed under certain conditions. Transactions in trade options are restricted to only those firms which handle the commodity in normal cash market channels. Firms offering trade option contracts would be required to become registered as agricultural trade option merchants (ATOMs) with the CFTC, to report their transactions to the CFTC, to provide their customers with risk disclosure statements, and to safeguard their customers' premiums. Also, these firms must meet the financial requirements of the CFTC.

Hybrid Cash Contract Problems

There have been numerous isolated incidents of delivery defaults and other problems associated with hybrid contracts, most notably HTAs and their variants. Most of these incidents result from a misunderstanding or a lack of disclosure of the potential risks that are inherent in these contracts. Another cause of hybrid contract problems is the inadequate structure and wording of some contracts which result in misunderstandings of the contract provisions by both parties. Particularly problematic is the separation of the control and financing/accounting functions that is present in the structure of most hybrid contracts. The producer has control over most of the pricing decisions of the contract, but it is the elevator that receives the account statements and is responsible for the margining of the futures and options positions. This separation of function can place a major strain on the financial resources of the elevator, forcing them to find other resolutions to the contract.

In response to the problems with hybrid contracts, the National Grain and Feed Association (NGFA) published a white paper titled "Hybrid Cash Grain Contracts: Assessing, Managing and Controlling Risk." A useful feature of this white paper is the inclusion of a risk matrix for each of the traditional and hybrid contracts which lists both the price and non-price risks present in each contract from both the seller's and buyer's perspectives. Also, the white paper contains sample contract language for HTAs that may minimize the potential for misunderstandings with these contracts.

Some variants of the HTA contract may be considered as illegal "off-exchange" futures or options contracts by the CFTC. To provide clarification to the regulatory atmosphere, in 1996, the staff of the CFTC published a set of "safe harbor" conditions on HTA contracts.

Issues Confronting Exchanges

The U.S. futures industry is in a state of transition as it deals with many of the structural changes in the U.S. and global economies. Changing world economies have increased the emphasis upon capital markets, particularly, securities and futures exchanges. In response, foreign exchanges have emerged and developed contracts that have been extremely successful, effectively ending the century-long hegemony that the U.S. exchanges have enjoyed. This added global competition has forced U.S. exchanges to reexamine the traditional methods of doing business.

The U.S. economy has experienced a prolonged wave of deregulation, starting with the transportation industries in the late 1970s, the telecommunications industries in the 1980s, and the natural gas and electric utility industries in the 1990s. This deregulation has opened opportunities and presented new challenges to the U.S. futures industry. Since deregulation exposes the industry to market forces, including price risk, new opportunities for futures contract development are emerging. Deregulation has also changed the overall structure of the traditional industries covered by futures contracts such as agricultural and financial markets forcing U.S. exchanges to rethink the traditional methods of conducting business.

In particular, these changes in the U.S. and global economies have brought to the forefront three major issues facing exchanges. These issues are contract delivery mechanisms, electronic versus open outcry trading, and product development.

Delivery Mechanisms[4]

One of the most important factors in the success or failure of any futures contract is the design of the delivery or settlement mechanism. It is the threat of delivery that keeps the marketplace honest. A well-constructed delivery mechanism provides an adequate arbitrage opportunity between the futures and underlying cash markets at contract maturity. The potential threat of this arbitrage is what forces convergence of the futures and underlying cash prices at contract maturity. If this convergence is a near certainty in the market participants' minds, then they must price the contract accurately to the underlying marketplace or ultimately fall victim to this arbitrage.

Traditionally, commodity futures contracts have relied upon delivery mechanisms based upon large defined centralized marketplaces. This worked well when food processing was concentrated in the major market centers such as Chicago, Minneapolis, and Kansas City. Regulated transportation rate structures emphasized these large "accumulation points" and as a result, most commodity futures emphasized the warehouse receipt for delivery of grains and the pen receipt for delivery of livestock.

With the passage of the Staggers Act of 1980, the railroads were allowed to set their rate structures competitively. The rate structures began to emphasize proportional rates instead of accumulation points. In addition, the elimination of regulated rates increased transportation costs for processed products versus raw commodities. This forced the relocation of the processing industries closer to the demand centers for the processed products, resulting in a more geographically dispersed food processing industry.

This decline of centralized marketplaces has resulted in more geographically disperse deliverable supplies for the some of the grain and livestock futures markets. Adding more delivery points to increase deliverable supplies only works when the geographic price distribution for the given commodity is uniform. However, if geography is an important aspect of the price, adding more delivery points has the effect of diluting the geographic basis of the par commodity and introduces substantial risk to the basis. Therefore, most commodity exchanges have been forced to consider other types of delivery/settlement instruments. These other instruments basically fall into three categories: cash settlement, origin certificates, and shipping (destination) certificates.

Cash settlements use an underlying cash market price or price index as a settlement instrument. The futures price converges by definition to the underlying cash price or index. The parties to a futures delivery do not deliver physical commodity, but rather settle through a cash payment from the buyer to the seller equal to the cash value of the futures contract. This method of settlement has been used extensively with financial futures. The first cash-settled commodity futures contract was the Chicago Mercantile Exchange's feeder cattle futures and options, which originally settled on the yellow sheet price index for live cattle. Because of geographic weighting problems with the yellow sheet index, the CME was later forced to construct its own settlement index for the contract.

Cash settlement works best for commodities where geographic definition is not an important aspect. For these commodities, one can construct an index of prices at such a wide variety of locations that any one entity cannot manipulate the settlement price. However, for commodities where geographic definition is important, designing an appropriate cash-settled index can be a two-edged sword. First, the index has to be broad enough to prevent manipulation. However, broadening the scope of the index reduces its geographic definition, thereby increasing the basis risk. Thus, cash settlement is generally not appropriate for geographically based commodities.

An origin (or throughput or loadout) certificate is a document which represents a claim upon a loading facility's (country elevator, etc.) ability to load out the par commodity onto a particular mode of transportation (rail and/or barge). The buyer immediately takes title to the commodity on-track or on-river and is responsible for transport to its ultimate destination. This is similar to an "on-track" purchase in the cash market. The load-out process is started when the buyer cancels his origin certificate. Furnishing transportation is usually the responsibility of the buyer, although making it the responsibility of the seller is also a possibility.

The futures contract using origin certificates for delivery will usually have specified a particular geographic region as "par" for pricing purposes. Facilities within this region can load out cars and/or barges and receive the par price. Other regions can be specified as "non-par" and have price premiums or discounts based upon freight cost differentials with the par region. Facilities are designated as "regular" by meeting the exchange's storage capacity, load-out rate, and financial criteria and are allocated a maximum number of outstanding certificates based upon physical and financial data provided to the exchange.

Origin certificates work well for commodities whose distribution of production is concentrated in small geographic areas. This delivery mechanism is less prone to "squeezes" and other delivery manipulations that can prevent futures-cash convergence; however, this depends upon the number of facilities that are registered as "regular" for delivery. Also, there is the possibility of a buyer's tying up the throughput capacity of a particular facility through the cancellation of certificates at inopportune times. This can be handled by setting up an additional tendering process at the time of certificate cancellation where the seller is randomly matched up with a particular facility for delivery.

Price convergence using origin certificates is a function of the potential deliverable supplies (i.e., the number of facilities listed as "regular") and the uniformity of prices across the par and non-par delivery regions. If prices are not uniform across the delivery regions, the geographic basis risk is increased. In addition, if the exchange is not precise in setting its premium/discount schedule for non-par delivery regions, it could create risk-free delivery arbitrage conditions that can distort the price convergence process.

Origin certificates are best suited to commodities where significant production is concentrated in small geographic areas and prices are relatively uniform across these areas. They are also well-suited for futures contracts on commodities that are slightly downstream in the marketing chain. The Chicago Board of Trade's soybean meal and the Winnipeg Commodity Exchange's canola and barley futures contracts are examples of commodity contracts using origin certificates for delivery.

A shipping (or destination) certificate represents a claim against a shipper's ability to provide conveyance (rail, barge, truck) to a designated par destination. This is similar to a "delivered" or "to-arrive" sale in a cash market. Upon cancellation of the shipping certificate by the buyer, the seller must load out a railcar, barge, or truck with tariff freight and switching charges prepaid to the contract's designated par destination. Each "regular" issuer of a shipping certificate is allocated a maximum number of outstanding certificates by the exchange. This maximum allocation is based upon the physical and financial capacities of the issuer.

Upon load-out of the mode of transportation and prepayment of tariff and switching charges, the buyer will forward billing instructions to the seller and a bill of lading will be drawn up for the account of the buyer. The buyer can then bill the railcars (or other mode of transportation) to their ultimate destination, and thus, absorb the freight differential between the par destination and their ultimate destination.

Shipping certificates are ideally suited to commodities whose production is widely distributed from a geographical perspective and where publicly quoted tariffs and switching charges are available. It is nearly impossible to "squeeze" a shipping certificate delivery. In addition, the deliverable supplies on a shipping certificate contract are equivalent to the entire production of the particular par commodity. Therefore, the threat of delivery arbitrage on a shipping certificate contract is very real and can be quite painful to those trying to manipulate the market.

One of the problems with the shipping certificate delivery mechanism is its dependence upon public tariffs. In the eastern United States, many of the major railroads use privately negotiated rather than published tariffs. This limits the eastern contracts, such as the CBOT's corn and soybeans, to barge-only shipping certificates. Another disadvantage to shipping certificates is that they have no intrinsic value due to a lack of backing by existing physical stocks as is the case with warehouse receipts. This also increases the possibility of delivery default since there is no verification of deliverable stocks before tender allocation on the contract.

The Minneapolis Grain Exchange uses shipping certificates for delivery on its soft white wheat, durum wheat, and feed barley futures contracts. The Chicago Board of Trade is proposing the introduction of barge shipping certificates on its existing corn, soybean, and wheat contracts to augment its existing warehouse receipt delivery mechanism. This change was necessary because of the closure of some significant regular warehouses in the CBOT's delivery markets and the subsequent loss of deliverable stocks.

An exchange does not necessarily have to limit itself to one of the delivery instruments as the only form of delivery. They can use combinations of these instruments, such as augmenting the warehouse receipt with shipping certificates as the CBOT is proposing. However, the risk in mixing delivery instruments is through the creation of uncertainty in the delivery process for the buyer. The buyer will not know the mode he will receive delivery until after tender allocation. This will incorporate a risk premium into the delivery price of the contract, which could be significant in size and hamper the price convergence process.

Electronic Trading

Foreign competition, much of which comes from electronic exchanges, has forced U.S. exchanges to consider electronic trading as either a substitute or as an after-hours segment for its traditional open outcry method of trading. The Chicago Mercantile Exchange, in partnership, introduced one of the first electronic trading systems called GLOBEX. In 1993, the New York Mercantile Exchange introduced its ACCESS electronic trading system. In 1994, the Chicago Board of Trade introduced its Project A system. All of these electronic systems were initially set up to facilitate after-hours trading of the exchanges' financial products, where competition from the big foreign exchanges is most acute. However, increasing competition in the commodity futures business has also led to the introduction of after-hours electronic trading of commodity futures.

Proponents of electronic trading cite the efficiency and convenience of placing orders. Also, since electronic trading takes place within computers rather than on a trading floor, there is less need for an exchange to create physical space to trade new contracts. In fact, many exchanges now tout electronic trading as an excellent method for introducing and developing new contracts.

Proponents of open outcry point out the difficulty with placing complex orders, such as multi-legged spreads, using electronic methods. Also, they point out that some valuable information is lost, such as the ability to gauge mood and reaction that comes from trading in the pit. Also, contributing to the opposition in the United States is the entrenchment of the pit trading culture and the fear of losing jobs to technology. This makes it very difficult for U.S. exchanges to introduce electronic trading for approval by their membership.

Almost everyone in the U.S. futures industry will accede to the opinion that eventually all of the U.S. exchanges must go electronic. The main point of debate seems to be how quickly should this change be implemented and what path should it take toward full implementation.

Product Development

Finding new futures and options contracts within the traditional commodity areas has become an increasingly difficult proposition for the U.S. futures exchanges. In addition, changes to the underlying markets have forced exchanges to reevaluate and change many of their flagship commodity contracts. Finding new market "niches" for contract development has become paramount in the priorities of the U.S. exchanges. Also, exchanges are looking for ways to diversify their product mix, just as in any business, to increase membership values.

However, introducing contracts into "niche" markets requires patience to let these markets develop. Rarely does an exchange find a "niche" market contract whose payoff, in terms of volume and open interest, is going to be immediate. In addition, developing a market "niche" may require that the exchange spend large amounts of resources to educate participants in the "niche" industry regarding futures and risk management.

In some cases, exchanges also face the "cart before the horse" problem when introducing any new futures contract. Locals want large commercial participation before they are willing to participate. However, the large commercials want market liquidity before they are willing to participate. Getting one or the other to jump in and "make" the market is the key to getting a new contract off the ground. Some exchanges try to "prime the pump" by using incentive programs that subsidize traders and commercials who trade the new contract. In some cases, these programs have been successful. In other cases, incentive programs have been guilty of creating high expectations and speculative bubbles, only to have volume and open interest come crashing down as the incentive money dries up and the program ends.

One of the problems larger exchanges face is that to some extent they have become victims of their own success. As memberships in the exchange become more valuable, the required rate of return or rental rate on the membership increases. Therefore, the holder of a membership will require a greater and more immediate rate of return for trading a new contract. If the contract is not an immediate success, the member's opportunity cost to trading the new contract increases, and they return to existing liquid markets. Some exchanges have tried to counter this effect by offering a special "class" of membership at a diluted value that only allows the holder to trade the new contracts. However, these reduced cost "memberships" often will face resistance from existing full members since these new memberships will have a negative effect upon the value of the full memberships.

To some extent, the smaller exchanges have an advantage in developing "niche" markets as compared to the larger exchanges. This may indicate that there is a unique role within the risk management industry for the various sizes of exchanges. For example, a small exchange, like the Minneapolis Grain Exchange (MGE), cannot bring to bear enough speculative capital to successfully trade a major financial contract such as the S&P 500 or the Eurodollar. However, the membership values of the MGE are low enough that it can continue to be patient with "niche" contracts such as its White and Black Tiger shrimp futures. The rental rate on an MGE membership is approximately $150 per month. Therefore, the holder of a membership can afford to occasionally go into the shrimp pit and place orders. At the CME, where the membership rents for $7,000 plus, members cannot afford to spend much time trading a shrimp contract.

New areas of commodity contract development include seafood, lumber, externalities (such as pollution rights), and energy. The energy area has explosive potential as the natural gas and electricity industries face deregulation. Electricity is potentially a $210 billion dollar industry with enough geographic peculiarities to warrant the potential for up to 16 regional futures contracts within the United States and Canada. Already, three exchanges (NYMEX, MGE, and CBOT) have electricity futures contract trading or proposed. Are there other areas for futures contract development? It is likely that the industry has only touched the tip of the iceberg.

Summary

There have been dramatic changes in the composition of futures trading in the past two decades. Most important has been the sharp increase in trading of all futures contracts within the United States, as well as off shore. However, it is notable that the growth has largely come from non-grain contracts. Within the grain sector, there have been sharp increases in all contracts, particularly since 1990. The contract with the fast growth rate has been the MGE HRS wheat contract.

Options on agricultural commodity futures began trading in October 1984 on an experimental basis and appear to be gaining in popularity. Introduction of options can be viewed as an innovation or as a new technology for marketing. Merchants or producers can use options in a multitude of ways, some as a complement and others as a substitute for traditional hedging. Their primary advantage for hedging is the ability to lock in a floor or ceiling price, and at the same time, allow the merchant/producer to take advantage of favorable price moves in the underlying commodity. The examples in this chapter demonstrate uses of options for long and short cash positions, and forward contracting.

During the past few years there has been an escalation in the number of different contract types now commonly used among traders and producers. These have been driven in part by competition within the grain industry, as well as demands by growers for more alternatives. In addition, these have been facilitated in part as a result of the existence of options. Finally, a number of these alternatives are on the fringes of challenging the scope of traditional regulatory regimes.

There are numerous issues that are confronting each of the United States grain exchanges. Most important are the structure of the delivery mechanisms, new product development and electronic trading. As these are resolved there is no doubt the conduct of the futures industry will change in the coming years.

Selected References

Belongia, Michael T., "Commodity Options: A New Risk Management Tool for Agricultural Markets," Federal Reserve Bank of St. Louis, St. Louis, MO, 1983.

Bowe, James, "Cutting Risk with Commodity Options," *Commodities*, December 1982, pp. 64-65.

Bullock, David, "An Introduction to Options Markets", Minneapolis Grain Exchange, Minneapolis, MN, 1993.

Chicago Board of Trade, *Options on Soybean Futures-Contracts Fundamentals, Pricing, and Applications*, Chicago, IL revised February 9, 1984.

_______, *Options on Soybean Futures-Fundamentals, Pricing, and Applications*, Chicago, IL, June 1984.

Chicago Board of Trade, "Options on Agricultural Futures", Chapter 1.

Chicago Board of Trade, "Protecting Deficiency Payments Using CBOT Ag Options, *The Commodity Futures Professional*, November 1987.

Crower, G., and E. Wailes. *Grain Marketing*, 2nd edition. Westview Press, 1993.

Delivery Issues Symposium, selected readings

Horner, David L., and Eugene Moriority, "The CFTC Options Pilot Program: A Progress Report," *Education Quarterly*, Vol. 35, No. 3, pp. 9-14, 1983;

National Grain and Feed Association, *Hybrid Cash Grain Contracts*, 1996.

Kenyon, David E., *Farmer's Guide to Trading Agricultural Commodity Options*, Agricultural Information Bulletin No. 463, USDA, Economic Research Service, Washington, DC, 1984.

______,*The Use of Futures Versus Put Options in Pricing Corn Production in Virginia*, Department of Agricultural Economics, Virginia Polytechnic Institute and State University, Blacksburg, VA, August 1984.

Klemme, Diana, "Ag Options, Missing Link in the Marketing Alternatives Chain?" Grain Storage and Handling, February 1984, pp. 26-33.

Klemme, Diana, "The Stanford Study and the Futures Market Delivery Process An Overview".

______, "Ag Options 2, A Merchandising Tool for the Country Elevator," *Grain Storage and Handling*, August 1984, pp. 29-45.

Mayer, Terry S., *Commodity Options: A User's Guide to Speculation and Hedging*, New York Institute of Finance, New York, 1983.

Minneapolis Grain Exchange, *The Power of Options*.

Moriority, Eugene, Susan Phillips, and Paula Tosini, "A Comparison of Options and Futures in the Management of Portfolio Risk," The Commodity Futures Trading Commission, *Education Quarterly*, Vol. 2, No. 1, 1983, pp. 5-11.

NGFA, pp. 5-11, and 29-33

Schwager, J., "A Complete Guide to the Futures Market," John Wiley and Sons, New York, 1984.

---

[1] Parts of this section are taken from the author's earlier chapter in Cramer and Wailes, pp. 229-236. The mechanics of options trading are explained in numerous publications including many directly available from the exchange.

[2] See Horner and Moriority and Moriority, Phillips, and Tosini for an early discussion of these issues.

[3] See NGFA for a thorough explanation of the multitude of different contracting forms and their hedging and risk management.

[4] See Klemme for description of conventional delivery mechanisms using warehouse receipts.

Wilson is a professor and Dahl is a research scientist in Agricultural Economics at North Dakota State University. Bullock is the chief economist at the Minneapolis Grain Exchange,

4

Grain Transportation and Grain Industry Structure

George W. Ladd

Effect of Technological Changes Upon Plant Size

Business innovations may be technological or institutional. The U.S. grain industry has experienced both. This chapter takes an empirical and a theoretical look at effects of technological change upon individual firms and upon an industry—in general, and in grain marketing in particular. To put the subject in some perspective, the chapter will first summarize French's conclusions about economies of scale.

Economies of Scale

French (1977, pp. 141-151) summarized the evidence on economies of scale from 98 different studies of cost functions of agricultural marketing plants. He concluded (p. 150):

> First, although substantial variability is observed, there appears to be some consistency in findings among studies dealing with particular types of commodities. For example, dairy processing plants and elevator and milling

operations, which involve larger inputs of capital items relative to labor, show greater scale economies than plants for egg or fruit and vegetable packing.

Second, plants that are, for example, only 20 percent of the size of the largest plant considered have costs from 10 to 40 percent higher than the largest plants for high labor operations and from 30 to more than 100 percent higher for the capital intensive operations such as dairy processing.

If we restrict consideration to in-plant operations, only two of the studies listed . . . show any diseconomies of scale within the size ranges considered. And even these cases. . .seem open to question. However, if we include assembly or distribution costs, the picture changes a bit. [Six] studies . . . show the optimum scale shifting back to the range of 40 to 80 percent of the largest plant, depending on the level of supply density or the transportation system. Several other studies . . . dealing with cotton, do not show the optimum scale shifting with the addition of assembly cost, but the magnitude of scale economies is reduced.

And finally, (p. 151):

The widespread existence of economies of scale, with the largest plant almost always showing the least cost, is clearly an important force in shaping the future development of the marketing system.

Economics of Scale: The 0.6 Factor Rule

One economic reason for the existence of economies of scale is the "0.6 factor rule" (Johnston, 1960 and Moore, 1959). The basic idea underlying this rule is that cost of many types of capital equipment varies directly with surface area but capacity is related to volume so that price, P, and capacity, C, are related by an equation of the form $P = \alpha C^{\beta}$ where the average value for ß is about 0.6. ß measures the elasticity of price with respect to capacity. That it is less than one indicates economies of scale in the purchase of capital equipment. A similar rule holds for construction costs. Increasing the length (or width) of a building by one percent increases the building's area by one percent. But it increases the circumference, and hence the wall length, by less than one percent.

The rule applies to grain handling facilities. Baumel *et al.* (1973, pp. 184-188) estimated installed annual costs of receiving, drying, storage, and load-out and cleaning facilities for different capacities at 1971–72 prices. The arithmetic mean value of ß from their data is 0.68. Cobia et al. (1986, pp. 95-98) presented estimates of 1984 capital investments for model elevators with different daily loadout capacities. The simple regression estimate of ß is 0.73.

Consider a firm that is planning to build new capacity (let $\beta = 0.7$). If it builds a plant having capacity of amount C, it spends $aC^{0.7}$ on purchase of new capital equipment; if it builds capacity of 2C, it spends $1.62\,(aC^{0.7})$. Doubling the amount of added capacity increases purchase cost by 62 percent. By building a plant with capacity of C it achieves an annual average fixed cost at capacity of $r(aC^{0.7}/C) = raC^{-0.3}$ where r converts purchase and installation cost to the annual fixed cost. If, however, it builds a plant having a capacity of 2C, its annual average fixed cost (AFC) at capacity is $r(1.62aC^{0.7}/2C) = .81raC^{-0.3}$, i.e., only 80 percent of the annual AFC for a plant of volume C.

Technological Change: Empirical

There has been one comprehensive study of effects of technological changes upon individual firms in an industry. In an extensive investigation of technological developments in dairy marketing industries, Williams *et al.*(1970), assisted by members of the North Central Regional Committee on Dairy Marketing Research (NCM-26), classified some 80 technological developments in the dairy industry into five categories according to the type of development and stage of marketing affected: (1) products, (2) processing and manufacturing, (3) packaging, (4) refrigeration, storage, and transportation, and (5) merchandising (pp. 35-45). They evaluated the technology's effects upon optimum volume of business, i.e., "the range in size over which unit costs of operation are relatively low" (Williams, *et al.*,1970, p. 36). Of 77 technological developments evaluated, they judged that 36 had a pronounced effect and 26 had a slight or moderate effect in increasing the optimum size of business. Only ten were judged to have had no effect; only five, to have reduced the optimum size of business. The great majority increased the optimum volume per plant by adding to capital requirements. Other effects that they attributed to some technological developments and that affected industry organization were facilitation of product differentiation and vertical integration, encouragement of brand advertising, and widening of plants' market areas.

They also compared several studies of costs of operating dairy plants (pp. 46-51). Their conclusion: ". . . technological change, by increasing capital requirements and fixed costs, commonly has increased the volume needed to operate most types of dairy processing and manufacturing plants at low unit costs" (p. 51).

Any innovation that increases a firm's optimum volume of business generates pressure on the firm to increase its actual volume of business.

The Williams *et al.* optimum volume is a range in physical volume about the volume of minimum average cost. Let the quantity at minimum average cost be q_{min}. They found that innovation increased q_{min}. Below q_{min}, average cost (AC) exceeds marginal cost (MC). At q_{min}, AC = MC. If price exceeds the value of AC at the q_{min} after an innovation is adopted (which exceeds the previous q_{min}) and the firm was operating in the neighborhood of the previous q_{min}, it has incentive to expand its operations; it must expand to maximize profit. See Figure 4.1 in which AC and MC are average total cost and marginal cost and q_{min} minimizes AC before adoption of an innovation and the primes (') denote costs and quantity after the innovation. The AC curves in Figure 4.1 do not possess the neoclassical U-shape; they are more J shaped to be consistent with economic-engineering cost studies, e.g. those cited by French, which almost always show that average cost declines until plant capacity is reached then turns sharply upward and increases rapidly with increasing volume. Because the AC curves turn up so sharply upon reaching their minimum point, a more descriptive label for them would be "sickle-shaped." In this Figure the firm maximized profit at q_{min} with the previous technology. After adopting the innovation, it maximizes profit at q'_{min}, and each increase in volume from q_{min} to q'_{min} increases its profit.

In some economic-engineering studies of short-run costs, the vertical part of the J-shaped cost curve (the "handle of the sickle") is implicit rather than explicit. Average costs are reported for volumes up to plant capacity. Output cannot be expanded beyond this volume with existing capital stock or can be expanded beyond it only by extraordinary measures.

There have not been any thorough studies in the grain industry of effects of technological developments upon individual plants' cost functions or optimum volumes of business. But observation of the grain industry leads one to believe that technological developments in grain handling have increased optimum plant size.

Technological Change: Theoretical

The work of Williams *et al.* provides substantial empirical evidence on the effect of technological change upon an individual plant. We will now use a simple firm model in a heuristic argument to gain some theoretical insight into effect of technological change upon an individual plant.

Figure 4.1 • Average Cost Exceeds Marginal Cost

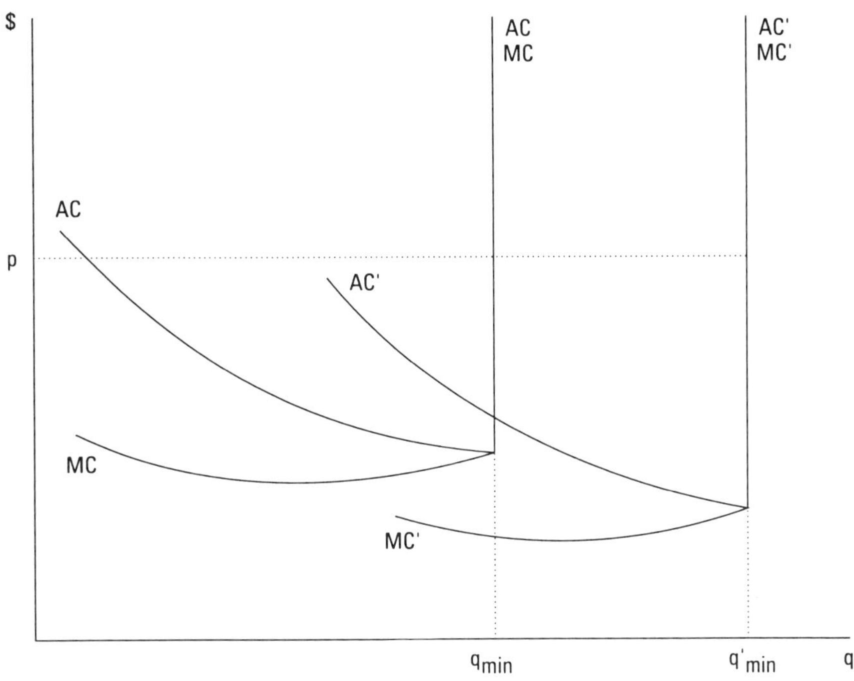

Let PC represent total cost of owning and operating a plant for one year, and let VPC and FPC represent annual variable and fixed plant costs. Let q represent the firm's physical level of operation. And we can represent the state of the arts, or technology, by t. Then we can write PC(q,t) and VPC(q,t) to mean that PC and VPC depend upon the firm's physical volume and technology, and can write FPC(t) to mean that the fixed costs depend upon the technology used by the firm. Likewise, let TC(q,t), VTC(q,t), and FTC(t) represent total transportation costs (assembly or distribution or both as the case may be), and variable and fixed transportation costs. It may seem artificial to treat t as "a measure of technology," but I believe we can reasonably do so if we interpret t as the date of introduction of an innovation to the market. Then a larger value of t denotes a more recent innovation. The firm's profit is $\pi = pq - (VTC(q,t) + FTC(t) + VPC(q,t) + FPC(t))$. Consider a firm that is in operation, and hence t is fixed. The profit-maximizing position is described by

$$\partial\pi/\partial q = p - (VTC_q + VPC_q) = 0,$$

$$\partial^2\pi/\partial q^2 = -(VTC_{qq} + VPC_{qq}) < 0.$$

VTC_q, VPC_q, and VTC_{qq} and VPC_{qq} are first- and second-order partial derivatives with respect to q.

Now, suppose the firm is making a long-run decision on the technology to use and the size of plant and transport facilities in a new plant. If it uses a technology different from the one it is now using, the effect on profit is

$$d\pi/dt = -(VTC_t + FTC_t + VPC_t + FPC_t).$$

The firm will adopt a different technology if $d\pi/dt > 0$. Typically, new technologies that are adopted reduce VTC and VPC and increase FPC and FTC: they substitute capital for labor. Hence, the firm will adopt a new technology if $-(VPC_t + VTC_t) > (FPC_t + FTC_t)$, i.e., if the savings in variable costs exceed the increases in fixed costs. (These same results would be obtained from Samuelson's (1947, pp. 57-69) model of cost minimization if he allowed fixed cost and optimum input levels to depend upon technology.)

Under what condition will adoption of a new technology increase the optimum level of output? The effect of adopting a different technology upon the firm's profit-maximizing level of output is

$$dq/dt = -(VTC_{qt} + VPC_{qt})/(VTC_{qq} + VPC_{qq}).$$

Because $(VTC_{qq} + VPC_{qq}) > 0$ by the second-order conditions, $dq/dt > 0$ if $(VTC_{qt} + VPC_{qt}) < 0$. Adoption of the innovation increases the optimum level of output if its adoption reduces the sum of marginal plant cost and marginal transport cost.

An innovation may affect costs and product quality simultaneously. Let $p_t = \partial p/\partial t$; p_t may be positive or negative because the innovation may raise or lower quality. Now

$$d\pi/dt = p_t - (VTC_t + FTC_t + VPC_t + FPC_t).$$

This is positive if $p_t - (VPC_t + VTC_t) > FPC_t + FTC_t$. The innovation increases profit if the change in price plus the savings in variable costs exceed the increase in fixed costs. And

$$dq/dt = (p_t - (VTC_{qt} + VPC_{qt}))/(VTC_{qq} + VPC_{qq}).$$

We have included the same t in plant costs and in transportation costs. We could use two different t variables. An alternative is to suppose that a technological change affects only plant costs or only transportation costs. In the first case $VTC_t = FTC_t = VTC_{qt} = 0$. In the second, $VPC_t = FPC_t = VPC_{qt} = 0$.

Market Area and Market Share

We see that we have reason to hypothesize that adoption of a new technology will increase a firm's most profitable level of output. This requires, obviously, that it increase its raw material purchases. These simple models have assumed that raw material supply is perfectly elastic. It is more reasonable to assume supply is less than perfectly elastic. Then the firm can increase its raw material supplies by paying a higher price. This will enlarge its procurement area or increase its market share within its procurement area. I now consider procurement area and market share for raw material, say grain. Consider two plants, one located at point one and one at point two. Let p_i be the f.o.b. plant price at plant i, d_{ij} be the highway distance from plant i to a farm located at point L_j, and t be the farmer's per-mile per-unit cost of transporting raw product to plant i. The net price at the farm gate for selling to plant i is $p_i - td_{ij}$.

First assume each farmer sells all of his output to one plant: to plant one if $p_1 - td_{1j} \geq p_2 - td_{2j}$ and to plant two if $p_1 - td_{1j} < p_2 - td_{2j}$. Each farm for which $p_1 - td_{1j} = p_2 - td_{2j}$ is on the border between the two market areas. For each farm j located on the border

$$d_{1j} - d_{2j} = (p_1 - p_2)/t.$$

The difference between the border firm's distances from the two plants is proportional to the price difference. Set $d_j = d_{1j} - d_{2j}$. It is clear that $\partial d_j/\partial p_1 > 0$. The effect of an increase in plant one's f.o.b. price is to increase d_{1j} and reduce d_{2j}, i.e., to expand plant one's procurement area and reduce plant two's area. A firm that has adopted a new technology has an incentive to increase the price it pays for raw product. Such an increase will enlarge its procurement area at the expense of its competitors. We also find that

$$\partial d_j/\partial t = (p_2 - p_1)/t^2.$$

Suppose $p_1 > p_2$. Then $\partial d_j/\partial t < 0$. A reduction in the farmer's per-unit per-mile cost of hauling product to a plant increases d_j; that is, it increases d_{1j} or reduces d_{2j}, or both. But, if $p_1 < p_2$, $\partial d_j/\partial t > 0$ and a reduction in t reduces d_{1j} and increases d_{2j}. In this model a technological change that reduces assembly costs of raw material increases the procurement area for the buyer that pays the higher price.

To focus on procurement market share rather than on market area, define

$$Pr(1,2,j) = \text{probability that a firm located distance } d_{1j} \text{ from plant}$$
one and d_{2j} from plant two sells its product to plant one.

$$= Pr$$

$$= \text{also proportion of raw product from sources located } d_{1j}$$
and d_{2j} from the two plants that is delivered to plant one.

Set $p_{ij} = p_i - td_{ij}$ and let $Pr = Pr(p_{1j}, p_{2j})$ where $\partial Pr/\partial p_{1j} \geq 0$ and $\partial Pr/\partial p_{2j} \leq 0$ and the first equality holds only at $Pr = 1$ and the second, only at $Pr = 0$. Now $\partial Pr/\partial p_1 \geq 0$ and $\partial Pr/\partial p_2 \leq 0$: a firm can increase its market share at the expense of its competitors by increasing the price it pays. Now $\partial Pr/\partial d_{ij} = (\partial Pr/\partial p_{ij})(\partial p_{ij}/\partial d_{ij})$, and $\partial Pr/\partial d_{1j} \leq 0$ and $\partial Pr/\partial d_{2j} \geq 0$: firm one's market share declines with increasing distance from firm one and increases with increasing distance from firm two.

To be able to say much about the effect of transport costs on market share, it is necessary to specify a functional form for $Pr(p_{1j}, p_{2j})$. First assume $Pr = Pr(p_{1j} - p_{2j})$. Then $\partial Pr/\partial t \gtrless 0$ as $d_{2j} \gtrless d_{1j}$, and both are independent of p_1 and p_2. An increase in t increases the probability of patronizing the closest firm.

If $Pr = Pr(p_{1j}/p_{2j})(= Pr(\rho)$, say), then

$$\partial Pr/\partial t = (\partial Pr/\partial \rho)(\partial \rho/\partial t) = (\partial Pr/\partial \rho)(p_1 d_{2j} - p_2\, d_{1j})/p_{2j}^2.$$

Because $\partial Pr/\partial \rho \geq 0$, it follows that

$$\text{sign } \partial Pr/\partial t = \text{sign } (p_1/p_2 - d_{1j}/d_{2j}).$$

If $p_1/p_2 > d_{1j}/d_{2j}$, then $\partial Pr/\partial t > 0$ and if $p_1/p_2 < d_{1j}/d_{2j}$, then $\partial Pr/\partial t < 0$. The firm with the relatively high price (high compared to d_{1j}/d_{2j}) loses market share as transport cost falls. This differs from the result we obtained when Pr depends upon the difference between p_{1j} and p_{2j}.

Each of the models in this section agrees with our intuition that a grain elevator whose costs are lowered by adoption of a new grain-handling technology can profit by increasing the price paid for grain in order to expand its procurement area or market share (or both) and thereby increase its grain volume. Our results on the effect of a change in grain-hauling technology are ambiguous, however, depending upon the functional form for Pr (1,2,j).

Capital Markets

The 0.6 factor rule, the existence of economies of scale, and the nature of technological innovation all favor larger firms. Their effects are compounded by the nature of capital markets. Paul (1964, p. 23) related cost of flotation of security issues to their size (see Figure 4.2) and reported that small firms:

> . . . incur higher costs because small security issues bear relatively high overhead charges and partly because they pay premiums to cover larger-than-average credit risks. The cost of raising funds by public means tends to rise precipitously with firms that grade downward from the level of the one million dollar asset size [in 1950s dollars].
>
> Firms with less than one million in assets [in 1950s dollars] pay from one to two percent higher interest rates when they borrow from bankers than large corporations. And when risks are judged too high, small firms might have to turn elsewhere or go without.

What Paul (p. 26) wrote about food retailing is probably true about all agribusiness industries:

> Large supermarkets do not ordinarily bring down unit costs. Presumably there might be about as much income produced by 50 independently owned supermarkets as by 50 supermarkets owned by one firm—assuming each group were serviced by comparable warehousing and buying arrange-

Figure 4.2 • Cost of Flotation Related to Size of Loan

Registered issues offered to the general public, 1951, 1953, and 1955

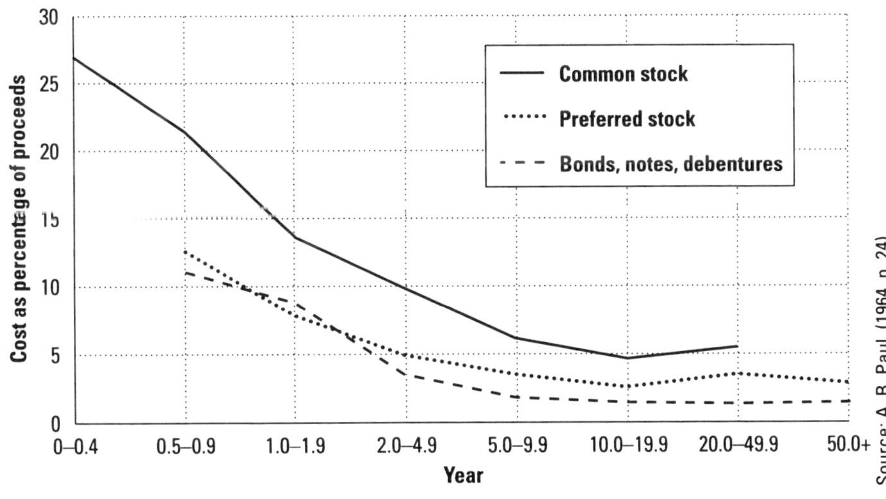

ments. In the chain, store managers are hired and the stockholders get the return on equity, whereas in the independent group, the store owners earn a management as well as an equity return. Conceptually, the debt-paying power of each group might be alike. But the 50 independents would have a serious financial disadvantage, which makes their very existence problematical. Outside investors may have legitimate doubts about the wisdom of a particular store site or reservations about the continuity of a particular management. In a 50-store chain, these doubts would fade because a few errors in location or weaknesses in store management would not ordinarily impair the firm's ability to make good on all its leases."

His insights apply to grain firms as well as food retailers, and we can validly replace "supermarkets" and "50-store chain" by "elevators" and "elevator chain." The risk a lender faces in lending a 20-elevator chain money to construct one more elevator is less than the risk faced by lending a single-elevator firm enough to add an equal-sized elevator. Likewise, the risk of bankruptcy faced by a multi-plant firm in borrowing money to build one plant is less than that faced by a single-plant firm in borrowing

Figure 4.3 • Average Total Cost without Shared Cost

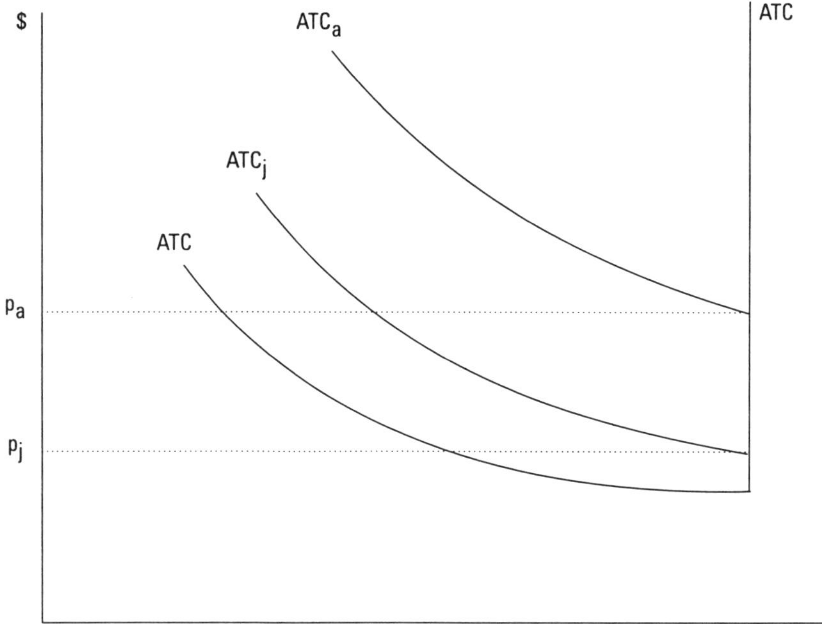

the same amount of money. This holds true whether the former firm's several plants are in different industries—or all in the same industry.

Paul's hypothesis has been amply confirmed by statistical comparisons of solvent with insolvent firms (or bankrupt with other firms). For example, in his study of solvent and insolvent grain elevators, McConnon (1989, pp. 148-149, 184-190) found that the probability that an elevator was insolvent was positively related to its debt-to-asset ratio. When all variables were set at their mean values, he reported that a 1.0 percent increase in the debt-to-asset ratio increased the probability of elevator insolvency by 0.2 percent. A given increase in debt increases the firm's debt-to-asset ratio and its probability of insolvency for a small firm by more than it does for a large firm. McConnon also found (pp. 148–149, 184–90) that the probability of insolvency was negatively related to cash flow-to-debt ratio. A 1 percent decrease in the cash flow-to-debt ratio increased the probability of insolvency by 0.4 percent when all variables were valued at their means.

Set-Up Costs for New Activities: Joint Ventures, Consolidation

Initiation of new activities usually forces a firm to incur fixed set-up costs as well as variable costs. A grain cooperative that has sold all its grain domestically and initiates international grain sales, or an elevator suddenly faced with substantial risk and a need to undertake risk management activities, both face set-up costs.

Consider an elevator manager in a period of stable, predictable prices when s/he has no need to use the futures market to hedge. Now let the source of stability be removed and let prices become highly variable and unpredictable. Now the manager has an incentive to take on a risk-management activity by using the futures market. Certain costs must be incurred whether the manager buys or sells one contract or one thousand. The elevator must invest in: (1) knowledge about measuring and responding to risk, (2) data sources, (3) data files, (4) computer hardware, and (5) computer software.

The maximum possible gain (or expected gain) is limited by the size (annual volume or capacity) of the elevator so long as the elevator uses the futures market only for hedging. The maximum gain for a sufficiently small elevator may be so small that it is not worthwhile or not even possible to incur the additional fixed cost. It may, however, be feasible and beneficial to merge or consolidate or undertake joint ventures with other small elevators to share the risk management activity and divide the fixed costs among several elevators. Suppose ATC in Figure 4.3 represents a firm's average total costs before undertaking a new activity; ATC_a represents average costs if it undertakes this new activity alone and must bear all set-up and variable costs; and ATC_j represents the firm's average costs if it undertakes the new activity jointly with other firms and shares the costs with them. At any price below p_a it cannot afford to undertake the new activity alone. But at prices between p_a and p_j it can afford to undertake the new activity jointly with other firms.

The 0.6 factor rule applies here, as does the existence of indivisibilities. The same equipment needed for one firm may be adequate for several firms, in which case each of the n cooperating firm's share of total fixed costs is TFC/n rather than TFC. And if larger equipment is needed for n firms, its installation and acquisition cost will increase less than proportionately with its capacity.

Some public policies have the effect of requiring firms to undertake new activities without increasing their volume of sales. Examples are environmental protection rules to obtain a cleaner environment and healthier citizenry that require firms to clean up their smoke or detoxify toxic chemicals, or OSHA's imposition of occupational safety or health requirements. It may be impossible to initiate these new activities with other firms. Because of the 0.6 factor rule, these requirements may be less onerous for large than for small firms because the large firm's increase in AFC at its volume of business will be less than the small firm's increase in AFC at its volume of business. Any firm that is covering its variable costs but not its fixed costs will be driven out of business because of its inability to afford the new undertaking.

J-Shaped Cost Curves Cause Excess Capacity To Be Desirable

Excess grain handling capacity—in a static sense—is inevitable and desirable in an economy in which there exists: (1) J-shaped short-run average total cost curves, (2) large and unpredictable variations in volumes of farm production, marketings, and exports, and (3) capital costs described by the 0.6 factor rule.

A number of observers have noted the existence of excess capacity in the U.S. grain marketing industry (Dahl, 1989, p. 117; Buschena, 1988 cited in Dahl, 1989, p. 117; Ginder, 1985, cited in Dahl, 1989, p. 118; Clow and Wilson, 1988, cited in Dahl, 1989; Dahl, 1990, p. 7; Hauser, Beaulieu, and Baumel, 1984; Adam and Anderson, 1985, p. 362; and Cobia et al.,

1986, pp. 7-8). Thompson and Dziura (1987, p. 120) found that in 1982 "Overcompetition and underutilization of capacity appears to be the case at many grain merchandising facilities in Illinois." Although Hanson, Baumel, and Schnell (1989); Baumel, Hanson, and Wisner (1987); and Hanson, Baumhover, and Baumel (1990b) did not refer to excess elevator capacity, their results indicated that elevators operated at below capacity and could reduce average cost of elevator operation by increasing volume. They found statistically significant evidence that increases in elevator utilization, measured by the ratio of storage capacity to annual shipments, resulted in higher bid prices to farmers for soybeans and wheat and larger margins to wheat elevators. They attributed the higher bid prices to economies of size in grain handling and the need to offer higher bids to attract larger volumes. It would not benefit the elevator to offer higher bids unless the increases in bids were offset by reductions in plant costs from the larger volumes. They also found statistically significant evidence that large corn elevators maintain larger margins than smaller ones.

The existence of J-shaped short-run average total cost (ATC) curves helps to explain the existence of excess capacity. Taken in conjunction, the J-shaped ATC curves and the unpredictable year-to-year variability in marketings and exports of farm products make the existence of excess capacity inevitable and desirable from the standpoint of having a low-cost system. (We here equate capacity with the plant volume that minimizes short-run ATC.)

Suppose that a plant's annual volumes are symmetrically distributed about volume q_2 in Figure 4.4, with minimum and maximum values of q_1 and q_3. Plant A, whose capacity level of operation is slightly greater than q_2, has the short-run ATC curve labeled AA. It achieves annual ATCs of c_1, c_2, and c_3 at annual volumes q_1, q_2, and q_3. Over a period of years, its mean value of ATC is C_a. Plant B has the short-run ATC curve BB. At volumes q_1, q_2, and q_3 its ATCs are c_1', c_2', and c_3' and its mean ATC over the years is c_a'. Plant B's capacity is slightly greater than q_3. Even though —in this figure— $c_1' > c_1$ and $c_2' > c_2$, c_a' is much smaller than c_a because c_3' is much smaller than c_3. The intersection of AA and BB is to the right of q_2. In most years, the smaller plant achieves a lower ATC than the larger plant. But the J-shaped ATC curve makes the penalty for operating above capacity so large that it is better (in terms of mean ATC) to have a plant that is so big that it never needs to be operated above capacity.

Now imagine a plant that has a U-shaped ATC curve that achieves its minimum at q_2. Its penalty—in terms of increased ATC—for operating above q_2 is little different from its penalty for operating at a smaller volume than q_2. Such a firm may economically operate above capacity as frequently as it operates below capacity because its penalty for exceeding capacity is small relative to the penalty for firms having a J-shaped curve.

In this connection, it is also worth recalling the point made earlier in connection with the 0.6 factor rule. Because of this rule, the construction of a larger plant permits attainment of lower average fixed cost at capacity than does construction and operation of a small plant.

In a world of unpredictably variable levels of farm production and of J-shaped ATC curves in agricultural marketing facilities, the continual existence of excess capacity is not convincing evidence of over competition, poor planning, over-optimistic forecasts, or poor firm management —although it may result from these. Rather, to have a system that minimizes mean ATC in a dynamic environment it is necessary to have excess capacity in most facilities most of the time.

Suppose that we interpret each J-shaped average total cost curve in Figure 4.4 as including all of an elevator's costs except its cost of grain; that we define "handling margin" to mean the difference between the elevator's selling and buying prices for grain; and that we take "required margin" to be the handling margin required to cover the average total costs (exclusive of grain, of course). The elevator's required margin is extremely large at low volumes, gradually declines until the minimum ATC is reached, and then increases rapidly as elevator volume rises still further. If the ATC curve is relatively flat for a large range of volumes below q_{min}, the required margin is nearly constant for a large volume range and increases rapidly as volume falls below or rises above that range. A substantial amount of excess capacity and a shortage of capacity lead to high required margins.

The grain industry was plagued by substantial excess capacity in the 1980s. This led to relatively high required margins, which exceeded actual margins for many elevators.

This graph is inadequate for explaining regional or national average handling margins because these depend on aggregate grain volume, not on an individual plant's volume.

This conclusion that excess capacity is less costly to a single elevator than a shortage of capacity applies to a situation of sizeable year-to-year variation in plant volume. Hilger, McCarl, and Uhrig (1977) studied a

Figure 4.4 • Average Total Cost Curves

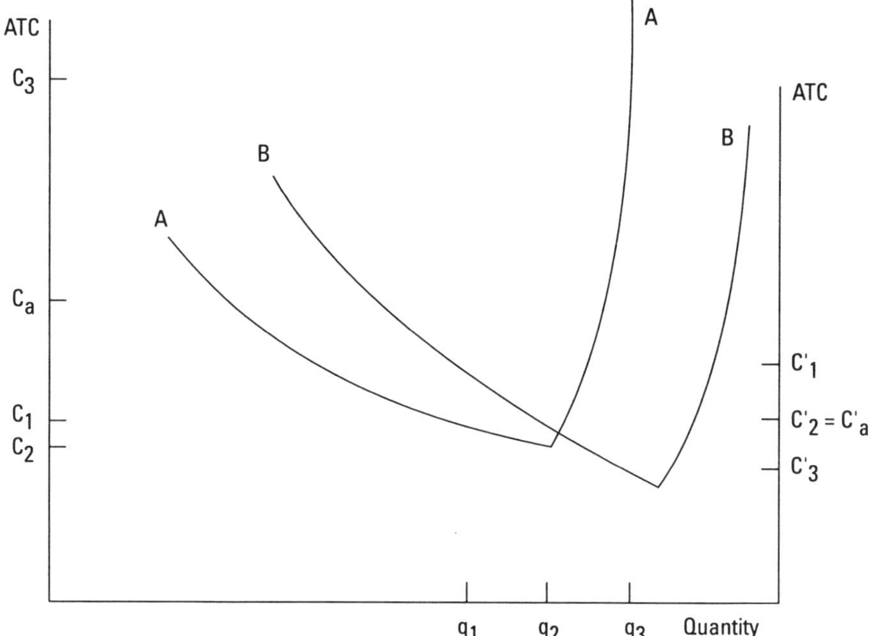

grain distribution system with fixed grain production. And they reported (p. 681) that here, too, overbuilding of subterminals in the grain industry was a less serious problem than underbuilding from a total industry cost standpoint.

Computer Models of Rail-Based Grain Distribution Systems: Unit Trains and Subterminals

There have been two waves of studies of rail-based grain distribution systems. The first wave addressed issues of adoption of covered hopper cars, introduction of unit-train rates, construction of subterminal elevators, and rail line abandonment. The second wave concerned the consequences of rail deregulation, mainly of the Staggers Act. This section covers the first wave. The next section will cover the second wave.

During the 1960s and 1970s the U.S. grain distribution system faced a number of dramatic changes. The first unit train tariff for grain was published in 1963, applying to wheat shipments from the Twin Cities in Minnesota to Buffalo, New York. In 1965 the Interstate Commerce Commission approved a unit-train rate for 44-car grain train shipments from Louisville, Kentucky to Charleston, South Carolina. The first multiple rail car rates on grain in Iowa were published in 1971. Unit train rates appeared later in the Great Plains states (Hauser, Beaulieu, and Baumel, 1984, p. 80; Baumel et al., 1973, p. 65). Baumel et al. (p. 4) summarized the changes in Iowa (which were typical of changes in other states also).

> In summary, recent innovations in grain harvesting and rail transportation, and changes in the supply of and demand for feed grains are some of the factors disrupting the grain distribution system. The production of corn and soybeans is increasing; larger volumes of grain are moving to more distant markets; new harvesting techniques are forcing huge quantities of corn and soybeans into elevator storage or market in short periods of time; railroad carriers are introducing multiple-car shipping rates, encouraging the use of jumbo covered hopper cars, reducing the number of 40-foot box cars, and proposing the abandonment of a significant proportion of track mileage . . . .

Then they went on:

> . . . neither the pricing system nor regulatory policies are adequately designed to coordinate or facilitate the industry adjustments needed to insure an efficient physical distribution system and provide for the general transportation needs of the grain industry.
>
> Innovations and changes in grain processing, transportation, and production are the source of many uncertainties and questions. Which rail lines should be abandoned? Where should grain handling facilities be located and how large should they be? And, what are the advantages of various grain distribution systems? Those who attempt to determine which rail lines should be abandoned often find that the location of subterminals must first be determined. Those who attempt to determine where subterminals should be located discover that it depends on the future of the rail network . . . and so it goes.

To obtain answers to these questions a number of studies of grain distribution systems were carried out in the 1970s: (1) Baumel and associates (Baumel et al., 1973, Ladd and Lifferth, 1975) studied a six-and-a-half county corn and soybean production area in north central Iowa; (2) Anderson, Gaibler, and Berglund (1976) studied a six-county corn and sorghum production area in south central Nebraska; (3) Salomone, Moser, and Headley (1977) investigated a corn and soybean area in northwest Missouri; (4) Larson and Kane's (1979) report covered a wheat, corn, and soybean area in central and southwestern Ohio; (5) Solomon, Larson, and Walker (1981)reported on a study of a wheat, corn, and soybean area in western Ohio; (6) Hilger, McCarl, and Uhrig (1977) studied a corn and soybean production area in northwestern Indiana; (7) Rudel and Lamberton (1976) included all grains in their study of the entire state of South Dakota; (8) Fuller et al., 1981, and Sorenson and Fuller, (1982) investigated hard winter wheat distribution from a 27-county area in Kansas, Oklahoma, and Texas. (i) Hauser, Beaulieu, and Baumel (1984) assessed the values of constructing additional multi-car grain handling facilities in various regions of Iowa and Nebraska under 1980 conditions. The results of these studies help us to understand the changes that the structure of the U.S. grain distribution system has experienced.

Not all studies addressed all of the issues identified in the preceding quotation from Baumel et al. For example, Larson and Kane studied only effects of rail line abandonment, and did not allow construction of new unit train facilities. By contrast, Sorenson and Fuller allowed construction of unit train facilities but did not consider rail line abandonment. The Iowa, Nebraska, and Missouri studies allowed for abandonment and/or upgrading of light rail lines, and for construction of new subterminal elevators and expansion of existing country elevators to subterminal status. The study in Indiana allowed for construction of subterminals, but did not consider rail line abandonment nor upgrading. The studies also differed in the quantitative method used: some used heuristic programs; two used network analysis; one, a decomposition algorithm. Some maximized grain producers' revenues; others minimized costs.

Sorenson and Fuller (1982) noted a number of significant differences between the hard winter wheat area they studied and the corn and soybean production areas (p.5):

> In contrast to the previous studies, railroad abandonment does not appear to be as imminent in the origin area of this study. Hence, all local elevators in this study are assumed to retain rail service throughout the analysis, whereas previous analyses included loss of rail service at selected local elevators.
>
> A second important difference is the presence of a major inland-terminal industry in the South Plains wheat-producing area, not present in the Midwest. Existing inland terminals have substantial grain handling and storage capacity, so that they do not require additional investment to accommodate unit trains. Substantial investments are required to upgrade local elevators into subterminals. This places potential subterminals at a cost disadvantage relative to inland terminals—a disadvantage that did not exist for subterminals in the Midwest.

A third difference is lower density of grain production, which is about one-fourth of that in the Midwest.

But there are enough common elements in these studies to enable us to draw some generalizations about effects of innovations in grain transportation upon the grain elevator industry.

Some of these generalizations follow.

1) Cited were earlier studies that concluded that almost all technological innovations in the U.S. dairy industry had increased optimum plant size. The adoption of covered hopper cars, unit train rates, and subterminal high-speed grain handling facilities likewise increased optimum grain elevator size.

2) Their adoption also reduced the minimum average total cost of elevator operation.

3) Rail line abandonment by itself increased total annual costs slightly (by less than 1 percent in the Ohio study by Larson and Kane (1979), which was only about 6 percent of the amount needed to upgrade and maintain the abandoned lines) and reduced grain producers' net revenue by an even smaller proportion.

4) But when rail line abandonment was combined with introduction of multiple-car rates and construction of subterminal elevators, costs fell (and producers' net revenues rose). This was true whether all light lines were abandoned or some abandoned and some upgraded.

5) Although elevators on abandoned rail lines were not forced out of business in the short run, they lost part of their grain business (some lost 80 percent or more), served mainly as storage facilities for grain marketed during harvest, and transshipped grain to market through subterminal elevators. Larson and Kane (p. 112) predicted, "They will probably add feed mill, farmstead equipment supply, custom fertilizer application service, and the like." It was likely that some of these elevators would be forced out of business in the long run because they would be at a serious competitive disadvantage relative to subterminals. Larson and Kane's prediction is supported by the results of Miller, Baumel, and Drinka (1977), who reported (p. 749) that Iowa elevators that had lost rail service had responded by searching for new product lines and improving their marketing practices, and also by eliminating unneeded labor and equipment.

One has to speculate on the reasons for the difference between the results from the computer models and the Miller, Baumel, and Drinka statistical results on the impact of loss of rail service on an elevator's grain volume. In the computer models, many elevators on abandoned lines lost business. Miller, Baumel, and Drinka found that Iowa elevators on abandoned lines had grown between 1969 and 1974.

a) It may be that the rapid growth in export demand for grain and soybeans during 1969–74 was sufficient to offset the potential loss due to abandonment. In the static computer models, this growth did not affect elevator volumes.

b) Part of the difference may be due to the nature of the implicit supply function to each elevator in the computer models: a stair-stepped function consisting of vertical segments (zero elasticity), horizontal segments (infinite elasticity), and corners. A change in elevator A's price from one cent below to one cent above elevator B's price causes a large amount of grain to be shifted from B to A if the change occurs at a corner. This sharp change likely overstates the farmers' actual response to price changes.

c) Results would also differ if elevators that lost rail service had previously competed mostly with other elevators that also lost rail service or if they had competed with elevators that did not lose rail service.

6) Country elevators that were on light rail lines that were neither abandoned nor upgraded would serve also mainly as storage facilities, transship most of their grain to market through subterminals, and would experience long-run competitive disadvantage.

7) Introduction of unit train rates and of covered hopper cars on heavy rail lines with continued use of boxcars on light lines, reduced long run grain assembly and distribution costs (and in creased producers' revenues) and stimulated construction of subterminal elevators.

8) Introduction of unit train rates and covered hopper cars along with abandonment of some light lines, upgrading some light lines to handle covered hopper cars, and construction of subterminals reduced costs and increased producers' revenues by more than did the steps in (7).

9) The greatest increase in producers' net revenues—by a small amount—in the Iowa study occurred when all light lines were abandoned.

10) Number of subterminals constructed was positively related to volume of grain exports and to the rate differential for unit trains. Reduction in the rate differential or the volume of exports reduced the optimum number of subterminals.

11) The later arrival (and later adoption) of unit train rates in the Plains states is reflected in the Hauser, Beaulieu, and Baumel (1984) finding that by 1980 shadow prices of multi-car grain loading facilities were substantially higher in Nebraska than in Iowa. They also found that several regions in Iowa had excess capacity in 25- and 50-car loading facilities but that there were large incentives for expanding multi-car grain loading facilities in Nebraska. A large number of multi-car facilities were constructed in Nebraska between 1980 and 1983.

Sorenson and Fuller (1982, p. 44) identified one side-effect of upgrading country elevators to subterminal elevators:

> For example, the current single-car rate allows management to merchandise wheat as it is purchased from producers and, therefore, purchased inventory is not great. In contrast, subterminal management would be required to accumulate inventory to meet the multicar shipment requirements of the unit train, and the risk associated with change in value of inventory created. This risk could be reduced through hedging with futures contracts.

The studies report savings in costs or increases in revenues for an entire system. For example, Hilger, McCarl, and Uhrig report (1977, p. 679) ". . . the [cost] difference between no subterminals and any subterminals is 1.25¢ per bushel indicating that subterminals offer significant savings in the grain marketing system." This is 1.25¢ for every bushel marketed commercially in the area. The reported figures almost certainly understate the cost savings and revenue gains obtained by the early adopters of subterminal capacity for handling unit trains.

One reason that few existing plants were driven out of business in these studies is that it is cheaper to operate existing facilities that are fully or partially depreciated than it is to replace them with new facilities. Of course, this raises the issue of replacing these existing facilities in the future. Will it be desirable to replace each existing facility as it wears out with a new facility in the same location? Of the same size? The answer will depend, inter alia, on location: Is the existing facility on a heavy rail line? It is liable to be advantageous—because of economies of scale—to replace several small worn out plants with one subterminal?

These studies of grain distribution systems were made during periods of increasing U.S. grain and oilseed production and exports and they assumed growth would continue. Between 1962–63 and 1972–73, U.S. corn and soybean production increased from 4.3 to 6.8 billion bushels and their exports rose from 0.54 to 1.5 billion bushels. By 1975–76, production and exports rose to 7.3 and 2.3 billion bushels. In the six-and-a-half county region that Baumel and colleagues studied, sales of corn and soybeans to processing and export markets amounted to 71 million bushels in 1970 and were projected to reach 118 million bushels by 1980 (Baumel et al., pp. 11, 14). In the six-county study area in Nebraska, commercial sales of corn and grain sorghum amounted to 39 million bushels in 1970 and were projected to reach 67 million bushels by 1980 (Anderson, Gaibler, and Berglund, p. 13).

The previously cited system-wide studies provide little insight into the effect of introduction of unit trains upon individual elevator pricing behavior. Schmiesing, Blank, and Gunn (1985) did study its effect. They collected daily corn price data for one year before and one year after establishment of a unit train facility by a local South Dakota elevator. Another unit train facility a few miles away from the case study elevator had opened a few months previously. They also collected daily cash price data for major regional and export markets and daily prices on the nearby futures contracts. They hypothesized (p. 95):

> The introduction of unit train technology causes fundamental changes in the way an elevator merchandises its grain. Using unit trains gives the elevator access to larger and more distant markets. Therefore, its ability to arbitrage between spatial markets increases. As a result, elevator managers may need additional methods to protect against adverse price movements on grain inventories held. Improved arbitrage is expected to cause linkages between cash and futures markets. In addition, to make effective use of unit trains the elevator must attract a larger volume of grain, which requires more competitive pricing by the elevator. In turn, the elevator would be expected to become more responsive to price changes at the major regional terminal markets.

Results in general confirmed their hypotheses. The local elevator's daily prices and daily price changes were much more highly correlated with prices and price changes at regional and export markets; and local daily price changes were more highly correlated with daily changes in the nearby futures contract price after the opening of the unit train facility than before.

Computer Models of Rail-Based Grain Distribution Systems: Staggers Act

The Staggers Rail Act of 1980 (P.L. 96-448, 1980) granted railroads the freedom to establish freight rates and provided various means of making upward rate adjustments, permitted them to enter into contracts with shippers and receivers, and allowed contract rates based on seasonal or other demand conditions. It also virtually eliminated the legal collusive rate setting power of rate bureaus. Hauser and Baumel (1987, p. 8) cited an Association of American Railroads survey of major grain-carrying railroads that found that 57 percent of the railroads' grain tonnage moved under contract rate agreements by mid-1985. Grain shippers feared that the result of such freedom would be substantial increases in rail rates on grain. Various economists constructed computer models of grain distribution systems to investigate possible effects, and likelihood, of rail rate increases on grain.

Miller, Baumel, and Narigon (1981) used a linear programming model to examine the effects of railroad rate increases on carriers, grain elevators, and farmers in two districts in Iowa: one about 90 miles from the Mississippi River, and one about 225 miles from the River. They found that: (1) Shippers that were most likely to absorb higher rail rates before shifting to a different transport mode or market were those that (a) used multiple car or unit train shipments, (b) shipped more than 70 percent of their corn and soybeans by rail, or (c) had relatively low railrates; (2) Demand for rail service was more elastic in the eastern region (-1.07) than in the western (-0.05 to -0.20); and (3) Railroad companies have more market power in areas farther from the Mississippi River (conditions (a) and (b) tend to make demand for rail transportation less elastic). They concluded by expressing their doubt that railroads would fully exercise their rate freedom under deregulation. Increases in rates by all railroad companies in the eastern district would substantially reduce railroad revenues and ton-miles. But increases in rates by all lines in the western district would increase rail revenues and profits. They also concluded no railroad could afford to increase its rates independently if it had competition from other railroads.

The purpose of the Koo, Thompson, and Larson (1985) study was to (p. 125) "... provide information ... regarding the likely structure of the grain transportation and marketing system in 1990." They included rail transportation capacity constraints for each state, barge capacity constraints for each river section, and storage capacity restrictions.

One set of models used average carrier costs to measure transport costs under the assumption that transportation prices equal average costs under deregulation in a competitive economy. Another set of models used average carrier costs for trucks and barges but used the 1979 rail rate structure to represent pre-Staggers charges for rail service. They found that national grain transportation costs would decline, with the greatest reductions in cost of wheat shipments, if transportation rate structures changed from the pre-Staggers rate system to a rate structure based on costs.

Fuller, Makus, and Taylor (1983) used a multi-commodity, multi-period, cost minimizing network flow model to determine the likelihood of rate increases on export-grain movements following the passage of the Staggers Act. The study assumed that intermodal rate competition is the only form of rate competition that limits upward rate changes by railroads. They included corn, soybeans, soft wheat, and hard and durum wheat. They altered railroad revenue-to-variable cost ratios between surplus production areas and ports to identify the maximum ratios that intermodal competition will permit railroads to attain. Rates or costs for competing transport modes were held constant. They determined the maximum revenue-to-variable cost ratios that could be levied in each region for each commodity before traffic was diverted from railroads to a competing mode. These maxima were then compared with actual historical ratios. The nature of the results for each commodity can be shown by quoting results for the hard and durum wheat producing regions (pp. 58-60):

> In general, the hard and durum wheat producing areas lack the transportation alternatives of other grain producing regions. This is reflected by the substantially higher revenue-to-variable cost ratios that railroads are able to attain in the Plains. In the eastern portion of the Plains, the truck-barge combination operating on the Arkansas and Missouri Rivers restricts revenue-to-variable cost ratios to a range of 1.1 to 1.6. In the eastern portion of the Northern Plains, trucks offer intermodal competition through movements to the Duluth-Superior port area. Analysis indicates reduced effectiveness of intermodal competition in the Great Plains (western Kansas, east-

ern Colorado, Texas-Oklahoma panhandles, western Nebraska, western Dakotas and Montana). Railroads can increase their cost ratio to nearly 2.0 before significant Great Plains' production commences to move by truck or the truck-barge combination, while in a portion of eastern Montana the ratio may be increased to 2.4 before wheat is truck-carried to the Columbia-Snake system. In the west central portion of the study region (eastern Colorado and Wyoming), intermodal competition permits railroads' revenue-to-variable cost ratios to range up to 2.5, the highest attainable by railroads on any export movement.

Model results indicate that intermodal competition is less effective in restraining rail rates (revenue-to-variable cost ratios) in the hard and durum wheat regions than the corn, soybean and soft wheat regions. Maximum ratios attainable by railroads range from less than 1.0 to as high as 1.7 in the corn, soybean, and soft wheat regions, with most ratios varying between 1.1 and 1.4. In contrast, maximum attainable ratios in the hard and durum wheat producing area range up to 2.5.

They found that the regional average maximum ratios of revenue to variable cost were nearly equal to the historic regional average ratios and concluded from their similarity that the pricing flexibility allowed by the Staggers Act would have little effect on producing regions' export rail rate levels.

Fuller and Shanmugham (1981) had previously used a network flow model to study intramodal (rail–rail) and intermodal (rail–barge or rail–truck and barge) competition. They minimized total annual costs associated with the handling, storage, and transportation of export-bound wheat from a 27-county area in Kansas, Oklahoma, and Texas. The short-run intermodal analysis showed that collective setting of rates would allow railroads to increase rate levels. But the long-run intermodal analysis found that there existed incentive to increase river and port facilities to increase barge movement and this would restrict railroads' ability to raise rates in all but the westernmost portion of the region studied.

Hoffman, Leath, and Hill (1985) analyzed the impact of seasonal rate variations, which were permitted for the first time by the Staggers Act. They aggregated corn, sorghum, barley, and oats into one "feed grain" in a linear programming, multi-time period, modified transhipment model to minimize the total cost of transporting, handling, and storing feed grains. The model's base solution—using constant rail rates—was used to estimate rail service demands by region and quarter of the year. From these, seasonal indices of demand were computed that were used in setting seasonal rate variations. In the first seasonal rate change scenario, only rail rates were changed. In the second, rail, barge, and truck rates were changed by the same percentage. The number of quarters in which rates were increased exceeded the number in which rates were decreased. They concluded (p. 92):

> A flexible rail-rate policy, with a 5 to 15 percent change in rates regardless of rate response by truck and barge, increased rail revenue, volume transported, and reduced seasonality of rail shipments, by creating incentives to reschedule rail shipments from peak to off-peak rail demand periods through the substitution of country storage for rail transportation. Thus demand for storage space should increase in most production areas during the peak rail demand periods. Transportation and storage costs generally rose which would imply a downward pressure on prices at the country elevator.
>
> An equal percentage rate response by truck and barge would seem most likely to occur because their revenue would generally be greater and their shipment seasonality would be less than without a rate response.

In the first scenario, however, larger variations in rail rates caused declines in total costs. The seasonal rate policy reduced seasonality of rail demand and also shifted it because storage in peak periods and later shipment during off-peak periods became economical. The authors pointed out that theirs was a short-run solution that did not allow for increases in storage rates and that the seasonality might have been less pronounced if storage charges had been allowed to vary.

Econometric Studies of Staggers Act Effects

Several investigators applied econometric techniques to actual grain or soybean prices and actual transport rate and cost data to determine effects of the Staggers Act. Some were "before and after" studies of time series data that used binary variables to divide the data into a pre-Staggers Act era and a post-Act era and to measure the Act's effects. A generic example of the equations they estimated is

$$y_t = \beta_0 + \beta_1 d_t + \beta_2 X_t + \beta_3 d_t X_t$$

where y_t and X_t are continuous time series, e.g., rail rate and rail cost index, and d_t equals zero in the pre-Staggers era and 1 in the Staggers era. The investigators do realize that such a procedure is unable to isolate the effects of the change in the legislative environment from other changes that occurred simultaneously. Several conditions other than the Staggers Act that affect transportation demand and supply occurred while the transportation and grain industries were adjusting to the Act, and it is not possible to allocate the values of β_1 and β_3 among these changes: (1) U.S. grain production declined; (2) foreign demand and exports weakened; (3) transport technology and costs were changing with increasing use of unit trains and construction of subterminals; and (4) there was a surplus of rail cars and barges due to expansion of their stocks under pre-1980 export markets and rates, and barge rates decreased in the early 1980s.

If the estimate of β_1 or of β_3 is significant, we can conclude that the relation between y_t and X_t is different since the Act than before, but we cannot be sure that it is different because of the Act. The difference may be due to changes (1) through (4).

In their study of the effect of the Staggers Act on rail rates for Kansas elevators, Babcock et al. (1985) found (p. 367) that price spreads between 14 Kansas local elevators and the Gulf rose by 37.3 cents per bushel between the first quarter of 1977 and 1981, and fell by 35 cents per bushel between the first quarter of 1981 and the first quarter of 1984.

They also studied characteristics of Kansas contracts for shipping wheat by rail. They found (p. 370) that the Kansas data were consistent with the hypothesis that ". . . large shippers have contract negotiation advantages not available to the small shipper." For example, the number of a shipper's contracts and the shipper's total storage capacity were highly positively correlated, as were number of contracts and number of locations from which a firm shipped grain. The smallest shippers in the study constituted 33 percent of all shippers, had 16 percent of the contracts, and operated in one state, whereas the largest shippers studied represented 24 percentof the shippers, had 41 percent of the contracts, and operated in 16 states on average. They wrote (p. 371):

> Many of the events of the past four years appear related to the exercise of competitive options that did not occur in the pre-Staggers period when (1) rail industry control was exercised through rate bureaus; (2) rail operating and pricing practices were more closely circumscribed by the Interstate Commerce Commission and (3) the newer options such as shipper/ carrier contracts were not available. Rail rate reductions appear to be responses to market conditions created by many events . . . . Deregulation did not create these conditions but it has contributed to a market environment that adjusts more quickly to changes than in the past.

They concluded that farmers had benefitted from reduced rail rates, many small shippers had adjusted well to the deregulated environment, rail deregulation had hurt some shippers, and that it might hurt others, especially small shippers who were dependent on a single railroad and had limited ability to negotiate rates.

Chow (1986) found reductions of slightly over one-third in published rail grain rates on the Central Plains (Eastern Colorado, Kansas, and Nebraska) during the five-year period immediately following deregulation, and speculated that (secret) contract rates are lower than published rates and their use would have led to larger rate reductions than the ones he reported. His speculation was confirmed by Babcock et al. (1985). In an examination of 127 rail contracts, they found that contract rates averaged 17 percent lower than the corresponding tariff rates (p. 371). Chow concluded that there was strong interrail competition in grain rates in the study area and that interrail competition was stronger for export than for domestic markets.

Fuller et al. (1987) used multiple regression to study monthly differences between port prices and farm level prices in each port's "associated hinterland region" for 1976–85. They reported that after deregulation: (1) export-rail rates in the Central and South Plains declined dramatically; (2) rates linking eastern Corn Belt states with East Coast ports declined modestly; and (3) rates linking Iowa and Illinois with Gulf ports declined, but this was due to declining barge rates rather than to deregulation. Their results—especially the contrast between (1) and (3)—agree with other evidence that rail regulation allowed cartel pricing. A cartel was effective in the Plains because of the absence of intermodal competition and ineffective in the Corn Belt because of its presence. They also reported their belief (p. 166) that "removing the immunity of rate bureaus and contracting have generated interrailroad competition in those areas of the U.S. where railroads have historically enjoyed monopolistic power and relatively high rates."

Adam and Anderson (1985) used an econometric model to study weekly corn and soybean bids of Nebraska country elevators from September 1978 to August 1984. They compared the pre- and post-Staggers Act eras and bids by elevators with and without contract rates. They found (pp. 363–364) "clear evidence of increasing level of elevator bids for corn and soybeans as well as increasing variability of corn bids in the aftermath of passage of the Staggers Rail Act of 1980. Part but not all of these effects appear to have been occasioned by the Act." They also decided that the pricing flexibility granted by the Act had increased competitive pressures among grain marketing firms. Possession of contracts allows elevators to increase bids (and may require it) and elevators without bids are forced to increase their bids. They wrote (p. 364) "Elevators having contract rates appear to have a rate advantage of up to 14 cents per bushel, which in the case of corn appears to be offset by competitors' bids."

Wilson, Wilson, and Koo (1988) studied rail-rate setting, rail and truck demand, and truck supply for hauling wheat from North Dakota to Minneapolis and Duluth. They estimated separate sets of equations for the pre- and post-Staggers eras, and identified three important changes. After deregulation rail costs became less important, and truck competition became more important, as influences on rail rates. The increased flexibility in rate setting allowed by deregulation resulted in an increased speed of adjustment of rates to changing conditions. They also found that the total elasticity of demand for rail services for hauling wheat—i.e., elasticity that included changes in trucking rates induced by changes in rail rates—changed from −0.54 to −1.06, a change that has significant rail-rate implications, as will become clear in the next section.

Hauser (1986) studied interregional and intertemporal changes in U.S. grain transportation rates from various regions by different modes to alternative export points. He observed that rate levels tended to increase prior to the Staggers Act and to decline thereafter. He found that rail rates responded to changes in barge rates under deregulation and that grain price relationships among ports affect grain transport competition and that there exists interregional intracommodity competition. For example, rates for shipping corn from the Great Plains exceeded rates for shipping wheat because of interregional competition with corn from the Corn Belt. This is one example of "nonmodal competition" (Hauser and Baumel, 1987, p. 6). Others are interregional (Iowa versus Nebraska corn) and interdestination (shipping Nebraska corn by rail to the gulf ports or to west coast ports). Thompson, Hauser, and Coughlin (1988) studied railroad rate-to-variable cost (R/VC) ratios for export bound corn and wheat before and after implementation of the Staggers Rail Act. They (p. 189) also found effects of interregional competition upon rail rates in regions distant from barge facilities. They did not find a clear effect of the Staggers Act upon rail rates.

The Staggers Rail Act was not the only legislation passed in the early 1980s. The Motor Carrier Act of 1980 (MCA80) greatly relaxed controls on entry into the commercial trucking business. The Surface Transportation Assistance Act of 1982 (STAA82) raised motor carrier taxes but also increased maximum truck weights, truck-trailer length, and trailer width. Babcock and German (1989) developed rail market share models. They found that (p. 251) "rail–truck market shares in the pre-deregulation period are a function of relative modal rates, interest rates, and relative modal service. However this formulation breaks down in the post-1980 period. Analysis indicates that post-deregulation time dummy variables accounted for little impact on rail–truck market shares in 1981 and 1982, but the effect steadily increased in most markets between 1983 and 1986. This could be due to lagged response to major transportation policy changes . . ." including MCA80, STAA82, and the Staggers Act.

MacDonald's (1987) study of competition among railroads did not address the issue of Stagger Act effects, but provides substantial insight into competition among railroads since its passage and helps to understand its effects. He studied rate competition among railroads for export shipments of corn, soybeans, and wheat. He measured competition from water transport as the mileage from the grains' origin point to the nearest location of water transport. Results for this measure were highly significant for all three commodities; rail rates increased as barge competition became more remote. He measured railroad competition within a region as shares of grain shipments by each railroad in the region; it was the reciprocal of a Herfindahl index. Its coefficients were also highly significant for all three commodities, indicating that increased rail competition—as measured by his concentration ratios—was associated with lower rates. He also investigated interactions between his measures of inter- and intramodal competition and concluded (p. 160), "Competition among railroads has a weaker effect on rates when there is nearby water competition, but becomes more important as one moves away from the water."

An origin contract is between a shipper and a railroad. A destination contract is between a grain buyer and a railroad. A grain buyer having a destination contract may use some of the contract rate savings to increase bids to elevators in order to attract the grain needed to meet the contract's volume requirement. Baumel and associates studied effects of rail contracts on grain elevator handling margins and on bid prices to farmers and elevators. They collected data for the years 1983–85 from wheat elevators in Kansas, Oklahoma, North Dakota, South Dakota, and Minnesota, and corn–soybean elevators in Nebraska, Iowa, Minnesota, and South Dakota (Hanson, Baumhover, and Baumel, 1990b; Hanson, Baumel, and Schnell, 1989; and Baumel, Hanson, and Wisner, 1987).

They found that destination contracts resulted in higher bids to elevators by buyers, higher bids to farmers by elevators, and higher grain elevator handling margins. All these results were statistically significant except for the positive effect on bids to farmers for wheat. Evidence on the effect of origin contracts was less consistent. Hanson, Baumhover, and Baumel concluded (p. 264), ". . . any origin contract rate saving received by the corn and soybean elevators are kept by the elevators and result in increased elevator margins. In the wheat elevators at least a portion of the rate savings from origin contracts are passed to farmers as increased price bids."

They also measured the effects of an elevator's ownership or leasing of rail cars by including a mileage and carpool allowance in the multiple regressions. They found that the allowances resulted in increased margins and bid prices to farmers for corn and soybeans; the effect on bid price to farmers for soybeans, however, was not statistically significant. They excluded this variable from the equations for wheat because it was highly correlated with the origin contracts variable.

Hanson, Baumhover, and Baumel (1990a) also used their sample of elevators to identify characteristics of elevators that had origin railroad contracts during 1983–85. They used a logistic function to measure the effects that five factors have on the probability that an elevator successfully negotiated a contract with a railroad: (1) annual total quantity shipped, (2) single- versus multiple-location, (3) number of railroads serving an elevator, (4) possession of branch elevator operations served by different railroads, and (5) existence of competing elevators located on competing railroads. They obtained statistically significant effects for factors (1), (3), and (4). Larger elevators—measured by total tons shipped in a year—are more likely to have a contract, as are elevators served by more than one rail line and multi-plant operations having plants on competing rail lines. The results for (3) and (4) imply that increased intramodal competition increases the probability an elevator can successfully negotiate an origin contract. One implication of the results is ". . . origin contracts encourage fewer but larger grain elevator firms" (Hanson, Baumhover, and Baumel, 1990a, p. 1045). This is a consequence of the contractual requirement to meet minimum shipment size and volume requirements. Destination contracting also favors larger elevators: it is cheaper for a buyer to sign two contracts with two large shippers of 100-car trains than with 20 elevators for 20 cars each.

Prior to the Staggers Act, rail rate bureaus were legal means for collective rate making. The Act removed the rate bureaus from antitrust immunity and increased the use of negotiated rates. Fuller, Makus, and Taylor (1983) concluded (p. 52) that "The Act would seem to virtually preclude legalized group rate making and increase independent pricing action." However, the small number of competitors in most regions and the knowledge of competitor behavior acquired during the long history of cooperative action in rate bureaus makes it likely that tacit cooperative rate setting is, or will be, practiced. Fuller, Ruppel, and Bessler (1990) cite several studies that found that the deregulation brought about by the Staggers Act led to rail rate reductions in the Central and Southern Plains by 1986. They held that inter-railroad pricing competition was facilitated by the legalization of contracting and the removal of rail rate bureaus from antitrust immunity.

As a consequence of small grain shippers' complaints that they were disadvantaged by the more favorable contract rates available to large shippers, Congress enacted Public Law 99-509 (21 October 1986) that requires disclosure of essential contract terms. See Fuller, Ruppel, and Bessler, p. 1046 for disclosure requirements. They found that the disclosure requirements had reversed the 1981–86 downward trend in rail rates. They concluded (p. 271), "This study suggests that contract disclosure and the increased reliance on posted tariff facilitated rate coordination by the oligopolistic railroad industry, thereby leading to an increase in rail rates" in the Southern Plains states of Kansas, Texas, and Oklahoma. "This finding supports the argument of grain shippers who contend that contract disclosure discourages interrail competition." They attributed the upward post-deregulation trend in rates to lack of effective intermodal competition and believe that there was less reason to expect the upward trend in the Corn Belt, where railroads compete with trucks and barges. Their belief is supported by Thompson, Hauser, and Coughlin's (1988, p. 193) results suggesting that barge competition has weaker competitive effects on R/VC (rate-to-variable cost) ratios in regions distant from water; and with Miller, Baumel, and Narigon's findings (1981, pp. 352, 357) that aggregate demand for rail transportation of corn and soybeans is elastic (less than –1) in eastern Iowa, close to the Mississippi River, and inelastic in western Iowa, farther from the Mississippi. In areas close to barge loading facilities, oligopolistic rail rates tend to follow competitive barge rates.

Fuller, Ruppel, and Bessler also wrote (p. 271), "The implemented disclosure policy represents a return to shipper rate equalization, a regulatory philosophy of the pre-Staggers era." One thing the implementation of a shipper rate equalization philosophy does is to balance the cost advantages achieved by larger shippers as a result of their economies of scale or economies of size.

Collusive Oligopolistic Pricing and Demand Elasticities

There is a simple relation between elasticity of demand and gross revenue for a single product firm or competitive industry. That relation is not so useful in studying oligopolistic or duopolistic pricing. Marginal revenue relates change in quantity to total revenue. What we will call "incremental revenue" may be useful for studying oligopolistic pricing; it equals the firm's change in total revenue per unit change in the firm's price. Intuitively, incremental revenue seems more appropriate than marginal revenue for the study of firms that are not price-takers because it measures the effect upon revenue of a change in the firm's decision variable.

Limited reading on rail transportation of wheat in the Plains states leads one to believe that the model to be presented does depict rate-setting behavior of railroads in these states during the pre-Staggers era and during the years since legislation has required publication of rail contract provisions. A modification might be appropriate for studying rate-setting behavior of rail lines close to the Mississippi River. The modification would have the entire rail industry close to the River being a price follower—following rates set by the barge industry.

Let p_i and q_i denote price and quantity of output of firm i, and p_j denote price of a duopolistic competitor (or average price of oligopolistic competitors). Suppose

$$q_i = f(p_i, p_j)$$

$$p_j = g(p_i).$$

The second expression describes interdependence of pricing decisions. Now the i-th firm's incremental revenue is

$$IR_i = d(p_i q_i)/dp_i = p_i[(\partial q_i/\partial p_i) + (\partial q_i/\partial p_j)(\partial p_j/\partial p_i)] + q_i.$$

Define cross-price elasticity of demand as $e_{ij} = (\partial q_i/\partial p_j)(p_j/q_i)$ and own price elasticity of demand similarly but with $j = i$. Also, let $(\partial p_j/\partial p_i)(p_i/p_j)$ = v_{ji}; v_{ji} represents firm i's price-conjectural variation, i.e., its conjecture about competitors' reactions to its price changes. I assume v_{ji} to be correctly known. Then incremental revenue can be expressed as

$$IR_i = q_i(e_{ii} + e_{ij}v_{ji} + 1).$$

If e_{ij} or $v_{ji} = 0$, we have

For inelastic demand ($e_{ii} > -1$), $IR_i > 0$ and $MR_i < 0$,

For elastic demand ($e_{ii} < -1$), $IR_i < 0$ and $MR_i > 0$.

For an oligopolistic industry, however, it is reasonable to suppose that e_{ij} and v_{ji} are positive. An increase in price by firm i results in increased sales by its competitors. We can take a positive value of v_{ji} as representing price leadership if firm i is a leader and j a follower, or as followership if firm i is a follower and j a leader, or as representing explicit or tacit collusion. Now, even though the demand for firm i's product is highly elastic (e_{ii} much less than -1), firm i may still increase (reduce) its revenue by increasing (reducing) price. This occurs if

$e_{ii} < -1$ but $1 + e_{ij}v_{ji} > -e_{ii}.$

And even though the firm's demand is inelastic (i.e., $e_{ii} > -1$), firm i reduces (increases) its total revenue by increasing (decreasing) its price if

$1 + e_{ij}v_{ji} < -e_{ii}.$

Following Wilson, Wilson, and Koo (1988, p. 322), we can define

$e_i^* = (e_{ii} + e_{ij}v_{ji})$

as total elasticity of demand for the i-th firm's output. Then we can express incremental revenue as

$IR_i = q_i(e_i^* + 1)$

and

$IR_i \gtreqless 0$ as $e_i^* \gtreqless -1.0.$

These relations are summarized in Table 4.1.

A monopolist's price and physical sales volume are negatively related: $\partial q_i / \partial p_i < 0$. A duopolist's price and sales volume may be positively related:

$(\partial q_i / \partial p_i) + (\partial q_i / \partial p_j)(\partial p_j / \partial p_i) \gtreqless 0.$

An elevator or railroad that has a monopoly can definitely increase its volume of sales by reducing price. If that same elevator or railroad is a duopolist, however, a reduction in price may reduce its volume of sales and its total revenue.

Table 4.1 • Relations Between Marginal and Incremental Revenues and Demand Elasticity: e_{ij} and $v_{ji} \neq 0$

Demand elasticity	Monopolistic firm (or competitive industry)	Oligopolistic firm[a]
Elastic $e_{ii} < -1$	$MR_i > 0$ $IR_i < 0$	$IR_i \gtrless 0$ as $1 + e_{ij}v_{ji} \gtrless -e_{ii}$
Inelastic $0 > e_{ii} > -1$	$MR_i < 0$ $IR_i > 0$	$IR_i \gtrless 0$ as $1 + e_{ij}v_{ji} \gtrless -e_{ii}$

[a]/ $IR_i \gtrless 0$ as $e_i^* \gtrless -1$.

Wilson, Wilson, and Koo found that the total elasticity of demand for rail transport for North Dakota wheat had changed from a mean of –0.54 in 1973–80 to a mean of –1.06 in 1980–83 (p. 335). A fall in the value of e_i^* from greater than to less than –1.0 has significant pricing implications. Because the fall in the value of e_i^* changes the sign of IR_i from positive to negative, the fall causes a variation in price to have an entirely different effect. When e_i^* exceeds –1.0 (i.e., $-1 < e_i^* < 0$) IR_i is positive and a rail rate reduction leads to smaller rail revenues. But when e_i^* is less than –1.0, IR_i is negative and a rate reduction leads to larger revenues. (The Wilson, Wilson, and Koo results apply to rail–truck competition in North Dakota. There appears to be no studies of rail–barge competition that have estimated conjectural variation or total demand elasticity.)

In this model, an oligopolist's loss of business from a price increase is small because business is gained back as other firms raise their prices. Likewise, an oligopolist's increase in business from a price decrease is reduced by the loss of business back to other firms as they reduce their prices. The existence of collusion—represented by the positive value of v_{ji}—reduces the market's responsiveness to duopolistic price changes.

The collusion exists because the duopolists' products are close substitutes: $e_{ij} > 0$. A consequence of the substitute relationship is to reduce the elasticity of demand for the industry.

Define e_{ij} and q_i and p_i as before, define $e = (\partial q / \partial p)(p/q)$ as the industry's elasticity of demand, where $q = \sum_i q_i$ and p is an average industry price. Assume $p_i = p_j = p$, and $dp_i = dp_j = dp$, i.e., all firms charge the same price. Then

$$e = \sum_i \sum_j e_{ij}\, q_i/q = \sum_i e_{ii}\, q_i/q + \sum_i \sum_{j \neq i} e_{ij}\, q_i/q.$$

Elasticity e is a weighted average of firms' own price elasticities and of cross-price elasticities. We know that increasing (decreasing) an industry's price does not increase (reduce) its total physical sales volume: $e \leq 0$. Hence $\sum_i e_{ii}(q_i/q) < - \sum_i \sum_j e_{ij}\,(q_i/q)$. The elasticity of demand facing the industry is less than the weighted average of the firms' own-price demand elasticities.

For simplicity, consider a duopoly. Then

$$e = (q_1 e_{11} + q_1 e_{12} + q_2 e_{21} + q_2 e_{22})/q.$$

If $e_{12} = e_{21} = 0$, then elasticity is

$$\hat{e} = (q_1 e_{11} + q_2 e_{22})/q.$$

Because e_{12} and e_{21} are positive, $e > \hat{e}$. The existence of substitute products within the industry reduces the aggregate elasticity of demand for the industry's output. If we assumed $dp_i = a_i dp$ and $p_i = b_i p$ and set $a_i/b_i = r$, then

$$e = \sum_i \sum_j q_i e_{ij} r_i/q.$$

This oligopolistic pricing model makes the case for secret price reductions interesting.

Suppose that the value of e_{ii} for a firm is less than -1.0 but its e_i^* exceeds -1.0 because e_{ij} and v_{ji} are positive. A price reduction by this firm results in a loss in revenue as competitive firms reduce their prices. If this firm could keep its price reduction secret from competitive firms but not from customers v_{ji} would be zero. Then e_{ii} and e_i^* would be equal and less than -1.0, and the price reduction would increase the firm's revenue. But it is probably not possible to keep a price reduction a secret from competitors while letting all customers know it. But now consider a customer that has a highly elastic demand for the firm's product. If the firm negotiates a secret price reduction to this firm alone, competitive firms cannot retaliate by reducing their prices. Hence the price reduction increases the firm's total revenue from this customer without reducing revenue from other customers.

Because there is a fixed cost component of negotiating and administering any contract —large or small— a stronger incentive exists to negotiate secret price reductions with large customers. (The discussion in section 9 on quantity premiums and discounts shows how the existence of a fixed cost component of marketing activities causes firms to prefer doing business with large suppliers. It can also be interpreted to show how the fixed cost component causes firms to favor business with large customers.)

Because grains and oilseeds are production inputs, I will also look at collusive pricing of inputs. Define w_{ti} and x_{ti} as price and amount of input t used by firm i. Firm i's profit is $\pi = pf(x_{1i}, x_{2i}, \ldots, x_{ni}) - \sum_t w_{ti}x_{ti}$.

The first-order condition (FOC) for profit-maximizing use of the u-th variable input is

$$\partial\pi/\partial x_{ui} = 0 = pf_u - (w_{ui} + x_{ui}\,\partial w_{ui}/\partial x_{ui}).$$

$\partial w_{ui}/\partial x_{ui} > 0$ because the firm must increase its price to attract more of the variable input. Hence

$$pf_u - w_{ui} > 0.$$

Define $e_{uij} = (\partial x_{ui}/\partial w_{uj})(w_{uj}/x_{ui})$ to denote cross-price elasticity of supply of input u to firm i in response to changes in price paid by firm j for this input. Setting $j = i$ defines e_{uii}: elasticity of supply to firm i in response to its own input price. And let $v_{ji} = (\partial w_{uj}/\partial w_{ui})(w_{ui}/w_{uj})$ denote the conjectural variation about price of input u. If firm i varies w_{ui}, the effect on its profit is

$$\partial\pi/\partial w_{ui} = (pf_u - w_{ui})[\partial x_{ui}/\partial w_{ui} + (\partial x_{ui}/\partial w_{uj})(\partial w_{uj}/\partial w_{ui})] - x_{ui}$$

or, using elasticity notation,

$$\partial\pi/\partial w_{ui} = (pf_u - w_{ui})(x_{ui}/w_{ui})\,(e_{uii} + e_{uij}\,v_{ji}) - x_{ui}$$

Suppose $v_{ji} = 0$. Then write the effect of variation in w_{ui} as

$$\partial\pi/\partial w_{ui})_v = (pf_u - w_{ui})(x_{ui}/w_{ui})e_{uii} - x_{ui}$$

If we subtract; allow e_{uij} and v_{ji} to be positive; and recall that $pf_u - w_{ui} > 0$ we find

$$\partial\pi/\partial w_{ui})_v > \partial\pi/\partial w_{ui}.$$

It is possible for $\partial\pi/\partial w_{ui})_v$ to be positive but $\partial\pi/\partial w_{ui}$ to be negative. That is, it is possible that in the absence of collusion or price followership, firm i could increase its profit by increasing its price paid and its usage of input u, but the presence of collusion means it will lose money by increasing the price it pays for the input. Then, paying a higher price under collusion simply means it will pay more for each unit of input u but will not acquire more units of input.

Obviously this discussion does not explain the history of rates in the 1980s because it excludes costs, but it illustrates how the incremental revenue concept might be used to study actual rates.

We can carry this one step farther by using the concept of nonmodal competition (Hauser and Baumel, 1987, p. 6). Reductions in barge rates and in rates of rail lines close to barge lines would put pressure on those rail lines not so close to barge lines to reduce their rates also to avoid losing markets for grain to the barges and their nearby rail lines. (See section 5 for more on nonmodal competition.)

Overriding Conclusion

The one conclusion reached in every study of grain transportation and that overshadows all others is that the strength of railroads' (monopolistic or oligopolistic) rate-making power—both pre- and post-Staggers—is an increasing function of distance from the two major U.S. river systems (Mississippi River and tributaries, and the Columbia River–Snake River system). For example, see Fuller, Makus, and Taylor's (1983) study of maximum rates (= revenue-to-variable cost ratios). They found that railroads could charge much higher rates in areas distant from these river systems without losing their business to alternative transportation modes than in areas closer to these river systems. Demand for rail transportation is less elastic in areas without barge or truck–barge competition. The evidence, then, is that rail lines can and do practice geographic price discrimination.

This ubiquitous result gains added significance from the Beaulieu, Hauser, and Baumel (1985) finding that for most of the country the direct demand for barge services is elastic, and from the Koo, Thompson, and Larson (1985) finding that national direct demand for barge transportation of grain is elastic (elasticity of about –2.4). This tends to impose a downward pressure on barge rates during times of excess capacity.

Below-Cost Pricing vs. Predatory Pricing

Connor et al. (1985, ch. 6) present three case studies of conglomerate behavior in the food manufacturing industries: of Phillip Morris–Miller Brewing Company, Procter and Gamble, and ITT–Continental Baking Company. Their conclusions agree with those of Marion et al. (1986). In their survey of studies of conglomerate firms in the grocery product industries, one of the conclusions of Marion et al. (1986) was (p. 254):

> Only a relatively small part of the conglomerate's business was committed to the product or geographic market selected for analysis. This meant that the conglomerate could pursue below-cost pricing or other strategies in these markets without threatening the profitability of its overall operations. This permitted pursuit of predatory conduct calling for forgoing profits in one period in anticipation of recouping lost profits after the market was restructured.

That last sentence suggests that it would be interesting to develop a simple model of predatory pricing. Marion et al. looked at predatory pricing in a situation comprised of: (1) a multi-product or multi-market, or both, firm that was the subject of analysis, (2) several competitor firms, and (3) several time periods. A model that contains all these dimensions cannot be simple. By abstracting from some of these elements, one can develop a simple model of predatory pricing. Before presenting it, We will respond to the reference to "below-cost pricing" in this quotation by presenting a simple model of loss-leader pricing (Holdren, 1968, pp. 40-42, 62-65, 126-128; Ladd, 1988). The similarities and differences between the two models may be illuminating and will show that below-cost pricing need not be predatory pricing. (Marion et al. did not assert their equivalence.)

Loss-Leader Pricing: Products

Below-cost pricing, what Holdren referred to as loss-leader pricing, can be profitable even though it has no effect on restructuring the market in the way that predatory pricing does. In fact, the maximization of profit may require a firm to lose money on some products. To put it another way, it may be that failure to lose money on one or more products reduces the firm's profits. This can be demonstrated with a static two-product model of an imperfectly competitive firm. Let p and q represent price and quantity of output, let subscripts 1 and 2 identify products one and two, and let $c(q_1, q_2)$ represent the firm's total cost. The firm's profit is

$$\pi = p_1 q_1 + p_2 q_2 - c(q_1, q_2).$$

One first-order condition for profit maximization is

$$\partial\pi/\partial p_1 = (p_1 - \partial c/\partial q_1)(\partial q_1/\partial p_1) + q_1 + (p_2 - \partial c/\partial q_2)(\partial q_2/\partial p_1) = 0.$$

After dividing by $\partial q_1/\partial p_1$ this can be written as

$$p_1 - \partial c/\partial q_1 = -q_1/(\partial q_1/\partial p_1) - ((p_2 - \partial c/\partial q_2)/\partial q_2/\partial p_1)/\partial q_1/\partial p_1.$$

The profit margin $p_1 - \partial c/\partial q_1$ may be positive, negative, or zero. For example, if $(p_2 - \partial c/\partial q_2) > 0$ and the two products are highly complementary ($\partial q_2/\partial p_1 < 0$ and large in absolute value), then it is possible for the profit margin $p_1 - \partial c/\partial q_1$ to be negative. We see that static profit maximization can require this firm to lose money on product one.

Of course, $p_1 - \partial c/\partial q_1$ is a special profit margin: price over marginal cost. Holdren was modeling retail grocery stores. He argued that grocery store marginal costs and average costs were equal and constant throughout relevant ranges of output. Ladd (1988) used a slightly different model. He defined total variable cost as $c_1 q_1 + c_2 q_2$ and allowed each c_i to depend upon levels of output of both products:

$$c_1 = h_1 (q_1, q_2)$$
$$c_2 = h_2 (q_1, q_2).$$

And he showed that $p_i \gtrless c_i - 0$ for $i = 1,2$. That is, at the output that maximizes profit, marginal revenue and marginal cost for each product are equal, but average cost for one product may exceed its average revenue.

This below-cost pricing is not the same as predatory pricing. One consequence of predatory pricing is the elimination of some competitors. In this model of loss-leader pricing, the number of competitors is constant. It is of course possible that a negative (or even positive but small) margin does reduce the number of competitors during the **next** decision period.

Loss-Leader Pricing: Inputs

The possibility of loss-leader pricing of inputs is relevant in considering markets for raw farm products. Let y_i and w_i represent amount of and price paid for input i (i = 1,2); $q(y_1, y_2)$ be the firm's production, and p be price of output. Profit is

$$\pi = pq(y_1, y_2) - c(q(y_1, y_2)) - w_1y_1 - w_2y_2.$$

One first-order condition for profit maximization can be written as

$$p\partial q/\partial y_1 - (\partial c/\partial q)(\partial q/\partial y_1) - w_1 = y_1/(\partial y_1/\partial w_1) - [p\partial q/\partial y_2 - (\partial c/\partial q)(\partial q/\partial y_2) - w_2] [(\partial y_2/\partial w_1)/(\partial y_1/\partial w_1)].$$

$p\partial q/\partial y_i$ is marginal value product of input i (MVP_i) and $(\partial c/\partial q)(\partial q/\partial y_i)$ + w_i is marginal outlay on input i (MO_i). This first-order condition can be written as

$$MVP_1 - MO_1 = y_1/(\partial y_1/\partial w_1) - (MVP_2 - MO_2)/(\partial y_2/\partial y_1)$$
$$\text{and } MVP_1 \gtrless MO_1 - 0.$$

The Holdren model explains below-cost product pricing. This model concerns "above-revenue input cost."

Holdren's model describes product-product cross–subsidization; this model describes input–input cross-subsidization. A firm that buys products from and sells production inputs to the same patrons can engage in input-product or product–input cross-subsidization. A grain elevator and farm-supply firm may find it profitable, for example, to treat fertilizer as a loss-leader in order to stimulate larger volumes of grain deliveries from patrons.

Predatory Pricing: Product

The word "after" in the last phrase in the Marion et al. quotation—"after the market was restructured"—calls for the use of a multiple time period model. We will use a two-period model. Marion et al. discussed predatory pricing in a multi-product or multi-market context. But it is possible to model predatory pricing in a single-product, single-market model. Now let p and q represent price and output of the firm's single product, subscript i represent time period i (i = 1,2), n represent the number of firms in competition with the one being modeled, $c(q_i) = c_i$ = total cost in period i, and d represent discount rate. The predatory-pricing assumptions are:

demand is $q(p_i, n_i) = q_i$

$\partial q_i/\partial n_i < 0$

$\partial n_2/\partial p_1 > 0.$

As n_i rises the number of buyers located close to the competitors' firms also rises and the time and travel costs for patronizing a competitor fall. Hence more buyers patronize competitors and $\partial q_i/\partial n_i < 0$. If competitors respond to a reduction in p_1 by reducing their prices, their profits in period one fall. If they keep prices the same, they lose customers and their profits fall. Marginal firms find their average revenue falling below their average variable cost (and possibly below average total cost) and hence they exit the industry. So $\partial n_2/\partial p_1 > 0$.

The present value of the predator's profits is

$$\pi = p_1 q_1 (p_1, n_1) - cq_1(p_1, n_1) + [p_2 q_2[p_2, n_2] - cq_2(p_2, n_2)]/d.$$

One first-order condition for maximizing profit is

$$\partial \pi/\partial p_1 = 0 = (p_1 - \partial c/\partial q_1)\partial q_1/\partial p_1 + q_1$$
$$+ (p_2 - \partial c/\partial q_2)(\partial q_2/\partial n_2)(\partial n_2/\partial p_1)/d.$$

$\partial q_1/\partial p_1$ and $\partial q_2/\partial n_2$ are negative, and q_1 and $\partial n_2/\partial p_1$ are positive.

Hence this equality can hold even if price exceeds marginal cost in both periods. To obtain some basis for understanding the implications of the predation assumptions, let us first ignore the predation assumptions by setting $\partial n_2/\partial p_1 = 0$. Let us also use e_i to denote price elasticity of demand for this firm's output in period i: $e_1 = (\partial q_1/\partial p_1) \, p_1/q_1$. The first order condition can now be manipulated to obtain

$$e_1 - (\partial c/\partial q_1) \, e_1/p_1 = -1,$$

from which we can obtain

$$p_1 = (e_1 \, \partial c/\partial q_1)/(1 + e_1).$$

If $e_1 < -1$, $p_1 > 0$. Assume $e_1 < -1$.

Now assume $\partial n_2/\partial p_1 > 0$. The last two products in $\partial \pi/\partial p_1$ can be rewritten as

$$(p_2 - \partial c/\partial q_2)(\partial q_2/\partial n_2)(\partial n_2/\partial p_1)/d = D/d, \text{ say.}$$

I have already argued that the product of $(\partial q_2/\partial n_2)(\partial n_2/\partial p_1)$ is negative. Assume $p_2 - \partial c/\partial q_2$ is positive. Then $D < 0$. The preceding first-order condition can now be written

$$e_1 - (\partial c/\partial q_1) \, e_1/p_1 = -1 - D/q_1 d.$$

The solution for p_1 is

$$p_1{}^* = (e_1 \, \partial c/\partial q_1)/(1 + e_1 + D/q_1 d).$$

Because D is negative, $1 + e_1 + D/q_1 d < 1 + e_1$ and hence

$$p_1{}^* < p_1.$$

A single-product firm operating in a single market may find it profitable to engage in predatory pricing in period one. And notice that this model does not even require the firm to operate at a loss in period one, but does allow it.

The preceding argument together with the first-order condition for p_2 provides some insight into the predator's pricing in period two, after some competitors have been eliminated and predatory pricing has reduced the number of alternative sources of supply to each buyer. Thus there exists fewer products to compete with (substitute for) the firm's product. A reduction in number of substitutes tends to make demand more inelastic; i.e., to increase e toward zero.

$$\partial e_2/\partial n_2 < 0.$$

The first-order condition for p_2 can be expressed in elasticity terms as

$$e_2 - (\partial c/\partial q_2)(e_2/p_2) + 1 = 0.$$

Hence

$$p_2 = e_2(\partial c/\partial q_2)/(1 + e_2).$$

Then

$$\partial p_2/\partial e_2 = (\partial c/\partial q_2)/(1 + e_2)^2 > 0.$$

Hence

$$(\partial p_2/\partial n_2) = (\partial p_2/\partial e_2)(\partial e_2/\partial n_2) < 0.$$

The predatory pricing in period one has reduced n_2, the number of competitive firms in business in period 2. This has made demand facing the firm less elastic. The consequence of forcing a reduction in n_2 is to permit an increase in p_2, the price in period two.

The model can be briefly summarized:

1) $p_1{}^* < p_1$: The fact that $\partial n_2/\partial p_1 > 0$ makes it profitable to lower price in period one.
2) Lowering price in period one reduces the number of competitors in period two.

3) Reducing the number of competitors in period two encourages the firm to increase price in this period.

Predatory Pricing: Input

Let y_i be amount of variable input purchased in period i at price w_i (i = 1,2), and write supply of input in period i as y_i (w_i, n_i) where n_i is the number of other buyers of the variable input. Assume $\partial y_i/\partial n_i < 0$: as the number of competitors bidding for this input increases, the number of suppliers located closer to competitors increases. Their time and money costs of selling to competitors declines and this firm loses patronage. Also assume $\partial n_2/\partial w_1 < 0$. As w_1 is increased, more competitors find themselves unable to cover variable costs and are driven out of business.

The present value of profit is

$$\pi = p_1 q(y_1) - cq(y_1) - w_1 y_1 + [p_2 q(y_2) - cq(y_2) - w_2 y_2]/d.$$

One first-order condition is

$$\partial \pi/\partial w_1 = 0 = [(p - \partial c/\partial q)(\partial q/\partial y_1) - w_1] \, \partial y_1/\partial w_1 - y_1$$
$$+ ((p - \partial c/\partial q)(\partial q/\partial y_2) - w_2) \, (\partial y_2/\partial n_2)(\partial n_2/\partial w_1)/d.$$

Define period i supply elasticity to this one firm as $E_i = (\partial y_i/\partial w_i)(w_i/y_i)$. If we assume momentarily that $\partial n_2/\partial w_1 = 0$, and use elasticity notation, this equation can be written

$$(p - \partial c/\partial q)(\partial q/\partial y_1) \, E_1/w_1 = 1 + E_1$$

or

$$w_1 = (E_1(p - \partial c/\partial q)\partial q/\partial y_1)/(1 + E_1).$$

Let D represent the second line of the right hand side of the expression for $\partial \pi/\partial w_1$, and let $\partial n_2/\partial w_1 < 0$. Then

$$w_1^* = \frac{E_1 \, (p - \partial c/\partial q)(\partial q/\partial y_1)}{1 + E_1 - D/y_1}$$

Now $D \gtrless 0$ as $(p - \partial c/\partial q) \, \partial q/\partial y_2 - w_2 \gtrless 0$. If $D > 0$ than $w_1^* > w_1$.

And predatory pricing leads to paying a higher price for an input in period one in order to eliminate some competitor firms in period two.

By manipulating the first-order condition for w_2, we can obtain

$$w_2 = E_2(p - \partial c/\partial q)(\partial q/\partial y_2)/(1 + E_2).$$

From this, if $p - \partial c/\partial q > 0$,

$$\partial w_2 / \partial E_2 = (p - \partial c / \partial q_2)(\partial q / \partial y_2) / (1 + E_2)^2 > 0.$$

As n_2 is reduced, the number of alternative buyers of the variable input declines and the elasticity of supply of the input to this firm falls. Hence $\partial E_2 / \partial n_2 > 0$, and

$$\partial w_2 / \partial n_2 = (\partial w_2 / \partial E_2)\, \partial E_2 / \partial n_2 > 0.$$

The predatory input-pricing in period one has reduced n_2, which in turn has made it profitable to pay a lower price in period two.

A pricing decision may be predatory in effect but not in intent. A loss-leader decision taken by a multi-product firm with the sole objective of maximizing short-run profit may have the additional effect of driving some competitors out of business. One reason that a product is a loss-leader may be that it attracts substantial patronage away from competitors.

Technological innovation can facilitate predatory pricing by early adopters. If the innovation allows lower average cost and the elevator, for example, can raise the price it pays for grain, competing firms that do not adopt the innovation for one reason or another—e.g., lack of capital— can be forced out of business.

Predatory Advertising

Marion et al. (p. 255) also concluded

> Evidence demonstrates that food conglomerates tend to prefer nonprice conduct made possible by advertising-created product differentiation. Not only will this tactic raise prices to consumers, but it may raise entry barriers as well.

This suggests developing a model of "predatory advertising" in which the period i demand curve would be $q_i(p_i, n_i, a_i)$, where a = advertising expenditure, and the assumption of predatory advertising would be $\partial n_2 / \partial a_1 < 0$.

Conglomerates

Although these models illustrate that predatory pricing may be profitable for a single-product firm operating in a single market, it seems intuitively reasonable to expect predatory pricing to be more common among multi-product firms operating in several markets. Such a firm may be able to combine loss-leader and predatory pricing. A low product price, for example, may increase current profit (loss-leader) and may also reduce the number of future competitors and allow higher future prices for several products. It is possible that a conglomerate would find predatory pricing profitable for a product and market where it would be unprofitable for a single-product firm operating in that market. As the predatory pricing reduces elasticity (of product demand or input supply facing the firm) it raises the possibility of profitable price discrimination.

Quantity Premiums and Discounts

It has been observed that companies prefer large customers over small and large suppliers over small. Two examples have been cited in this paper. The earlier section on capital markets cited Paul's (1964) finding that small security issues incur higher average costs than large issues. The section on contracting presented Hanson, Baumhover, and Baumel's (1990a) finding that grain contracting favors larger elevators. MacDonald (1987, p. 28) presents several reasons that railroads can realize lower average costs with larger shipment sizes than with small sizes. This section presents a simple model that describes such behavior. The description has two major components.

1) The cost of acquiring an input exceeds the outlay on the input. A fixed cost of the amount a must be incurred for each supplier from which input is acquired, whether the supplier provides 1 percent or 90 percent of the firm's needs. This, naturally, causes the firm to favor larger suppliers. The value of a represents the minimum cost of record-keeping, check-writing, contract negotiation, quality inspection, etc., that must be incurred.

2) The first-order conditions use generalized optimization, which explicitly imposes the restriction that choice variables be non-negative.

Define

q_i = quantity purchased from each supplier in size group i; fixed

n_i = number of suppliers in size group i that are patronized

TC_i = total procurement cost for buying quantity q_i from firm in size group i

= $a + c_i q_i$ for $n_i > 0$

= 0 for $n_i = 0$

a = total fixed cost for patronizing a supplier

c_i = average variable cost for patronizing a supplier in group i

Assume the firm wants to minimize the total cost of procuring a fixed amount, Q, of an input. The problem is to select non-negative values of n_i to minimize

$$\alpha \sum n_i + \sum n_i c_i q_i = c$$

subject to

$$\sum n_i q_i \geq Q, \; n_i \leq \bar{n}_i$$

where Q is the total amount needed and $\bar{n}_i$ is the number of suppliers in size group i. The Lagrangean expression can be written

$$L = a \sum n_i + \sum c_i n_i q_i - \sum \mu_i (n_i - \bar{n}_i) - \mu_{n+1} (\sum n_i q_i - Q).$$

We note that: (1) $\mu_i = \partial c / \partial n_i \leq 0$, i.e., an increase in the number of firms in size group i cannot increase total procurement costs; and (2) $\mu_{n+1} = \partial c / \partial Q \geq 0$, i.e., increasing the required volume cannot reduce total procurement costs. One set of first-order conditions (FOC) is

$$\partial L / \partial n_j = a + (c_j - \mu_{n+1}) \, q_j - \mu_j \geq 0 \text{ for } j = 1,2,...n \tag{1}$$
$$n_j \, (\partial L / \partial n_j) = 0.$$

Another set is

$$\partial L / \partial \mu_j = \bar{n}_j - n_j \geq 0 \tag{2}$$
$$\mu_j (\partial L / \partial \mu_j) = 0 \quad j=1,...,n.$$

And the last FOC is

$$\partial L / \partial \mu_{n+1} = Q - \sum n_i q_i \leq 0 \tag{3}$$
$$\mu_{n+1} (Q - \sum n_i q_i) = 0.$$

Assume $n_j = \bar{n}_j$ (i.e., the firm does business with every supplier in size group j) and $\mu_j < 0$. Because (1) is now satisfied as an equality

$$a + (c_j - \mu_{n+1}) q_j = \mu_j.$$

Assume volume requirements are exactly satisfied and $\mu_{n+1} > 0$. Then

$$(a - \mu_j)/q_j + c_j = \mu_{n+1}. \tag{4}$$

This equality holds for any size group from which the firm acquires all the group's raw material. Now suppose that $n_t = 0$. Then $\mu_t = 0$ from (2) and

$$a/q_t + c_t > \mu_{n+1}. \tag{5}$$

This holds for each size group t from which the firm obtains no raw material. Because $\mu_j < 0$, it follows from (4) and (5) that

$$a/q_t + c_t > a/q_j + c_j. \tag{6}$$

a/q_j and c_j can be interpreted as average fixed cost and average variable cost of patronizing a supplier in size group j. So (6) can be written as

$$AC_t = AFC_t + AVC_t > AFC_j + AVC_j = AC_j.$$

The average cost of patronizing firms with which it does no business is greater than the average cost of patronizing firms with which it does do business. Suppose also that $c_t = c_j$. Then from (6) we have that

$$q_j > q_t.$$

The suppliers with which the firm does business are larger.

Now let us compare size groups j and s when μ_j and μ_s are negative and $c_j = c_s$ and $q_j > q_s$. Then we can find that $\mu_j < \mu_s$. That is, $\partial TC/\partial n_j < \partial TC/\partial n_s$; an increase in the number of large suppliers reduces total procurement cost by more than an increase in the number of small suppliers does.

A firm may also allow quantity discounts (to customers) or quantity premiums (to suppliers) as a means of predatory pricing. A firm may increase its competitors' average procurement or marketing costs by bidding large patrons away from them. To obtain any specified volume—Q— the competitors must then do more business with small patrons.

Earlier it was suggested that the formulation of oligopolistic pricing decisions provided some insight into secret price reductions to selected customers. This present discussion presents reasons for a firm to prefer negotiating secret selling price reductions with large customers (or secret procurement price increases with large suppliers).

What Is A Low Cost, Efficient Grain Merchandising System?

One purpose of economists' studies of marketing systems is to judge their performance: to evaluate their social, economic, and political consequences. Various writers have described the U.S. grain handling system as a low cost system. What they seem to mean is that the system has low private, pecuniary unit cost of production. Others have described the U.S. system as an efficient system, and some have compared the efficiency of the U.S. system with other systems. These studies equate efficiency with low private, pecuniary unit cost of production (see Lang for the economists' meaning of economic efficiency). Others describe the U.S. system as "low cost, efficient." They seem to distinguish low cost from efficient, though what they mean by efficient is never clear.

Before studying efficiency of the U.S. grain marketing system, we need to recall what French (1977, p. 94) wrote:

> The definition and dimensions of efficiency vary at different levels within the market economy and become increasingly complex as we move from the firm to an industry or group and finally to the total system.

I have argued (1983, p. 2):

> I have strong objections to taking the same accounting rules that proprietary firms use to measure pecuniary costs and revenues and using the rules alone and unaltered in public policy studies. Part of the job of such research is to measure outputs and inputs, costs and benefits beyond the ones that businessmen consider.
>
> Suppose Cargill owned the Mississippi River. Then the fish and bird habitats would have economic value because sport fishing and bird-watching rights could be sold. In a study of efficient grain transportation systems, we would have to take account of any loss of fish and bird habitats due to barging and dredging, and of the resulting loss of revenue to Cargill. But the Mississippi River is not privately owned. So the economic value of fish and bird habitats are not defined. So economists frequently ignore the effect of barge traffic upon dredging and the effect of piling dredge spoil upon bird- and fish-breeding areas.
>
> Now please don't tell me that "enjoying fish and wildlife is only a sentiment." Desire to maximize profit is "only a sentiment." So is desire to maximize utility. So is love.
>
> I argue that efficient behavior of an economic agent cannot be identified without a knowledge of the agent's objectives. Because we do not know society's objectives, we cannot identify efficient public policies.

A given grain distribution system can be both efficient and inefficient: efficient according to one criteria and inefficient according to another. In Ladd and Lifferth's (1975) study of grain distribution systems, the objective was a system that maximized net revenues of grain producers. Hilger, McCarl, and Uhrig's (1977) objective was to develop a system that minimized grain distribution costs. These studies used different criteria to determine efficiency. A system that was efficient according to the Ladd and Lifferth criterion was inefficient according to the other, and vice versa. Applying my position leads me to conclude that we cannot test a marketing system's efficiency until we have specified goals or objectives. And after we have done that we can argue whether an efficient system is a desirable one and which "efficient system" is "really more efficient." Ladd and Lifferth, for example, believed that their efficient system was more desirable than Hilger, McCarl, and Uhrig's efficient system. Authors who tell us that "a system is efficient" tell us nothing unless they also identify the criteria they used.

We usually assume there is a close relationship between efficiency and costs—by which is usually meant private, pecuniary, unit costs of production (Lang, 1980). But we have an inadequate measure of performance unless we consider social costs.

The works of the public-finance economists on merit wants, or merit goods, (see, Musgrave 1959, 1986 or Pazner, 1972) and Lang's paper on efficiency of public policy are relevant to this discussion. Public education, nutrition, housing are examples of merit wants; cigarettes and addictive narcotics are examples of merit bads. In an economy having no merit goods or merit bads, the Pareto Efficiency condition that maximizes a social welfare function equates every consumer's marginal rate of substitution (MRS) between every pair of goods to the community's marginal rate of transformation (MRT) between them and such an outcome can be achieved by a competitive price system. In an economy that has merit goods or bads, equating MRS and MRT fails to maximize a social welfare function and ". . . unfettered reliance on the competitive price mechanism cannot lead to an optimal outcome" (Pazner, p. 468); public intervention is required.

To put it another way, if every pair of MRS and MRT is equal in an economy that has merit goods or bads, the economy is inefficient because its behavior is inconsistent with its goals.

"Efficient" then, means no more than "efficient according to the specified objective." It also means "efficient subject to constraints imposed," as was clearly shown by Lang. He considered imposition of a national policy that restricted the use of livestock feed-additives because their use creates risks to human health, and of a policy that limited the size of farms to 160 acres if they used water from Bureau of Reclamation dams. He showed that, even if such policies were to increase the private pecuniary costs of agricultural production, the policies would not create economic inefficiency. A Pareto Efficient allocation of resources can be achieved after the imposition of either restriction, as well as before. The two efficient allocations are different but economic criteria are not sufficient to identify the preferred or superior allocation. Extra-economic, i.e., non-economic, considerations are required in order to compare Pareto Efficient allocations.

Because "efficiency" means "efficiency according to the specified criterion" and because cooperative firms do not have the same goals as proprietary firms, I do not believe it is meaningful to use the same standard to study efficiency of cooperative and proprietary firms. We usually assume that proprietary firms have the goal of maximizing profit. This cannot be the goal of a cooperative because a cooperative has no profit. Royer (1978), and VanSickle and Ladd (1983, 1986) assumed a cooperative's goal was to maximize members' profits. They found the first- and second-order conditions for efficient cooperative behavior to be different form the conditions for efficient proprietary behavior. Also Ladd (1974) demonstrated that an efficient farmers' bargaining cooperative that maximizes member price for members' (single) raw material behaves differently from an efficient bargaining cooperative that maximizes the volume of raw material marketed under its management. A cooperative that is operating efficiently according to its own standards, i.e., is behaving in the best way to achieve its own objective, will appear inefficient if tested according to the standards that are appropriate for a proprietary firm.

We have never addressed the question of devising appropriate measures of efficiency for cooperative firms. We have uncritically applied measures developed for proprietary firms. I believe this choice is an error. Consider, for example, measurement of a single-product firm's degree of short-run efficiency in achieving minimum expenditure on the variable inputs needed to produce a specified level of output. Developing such a measure for a proprietary firm involves minimization of $\sum w_i x_i$ subject to the requirement that $f(x_1, x_2,..., x_n) = \bar{q}$, where $\bar{q}$ is a fixed quantity of output, w_i and x_i are (fixed) price and amount used of the i-th variable input, and $f(\cdot)$ is the firm's production function. The familiar result is that input price ratios equal marginal rates of substitution:

$$w_i/w_j = f_i/f_j \tag{1}$$

where $f_i = \partial f(\cdot)/\partial x_i$.

(Parenthetically it is worth noting that expression (1) is also obtained by maximizing q subject to the production function and to the requirement that $\sum w_i x_i$ is fixed.)

I doubt that this formulation is even relevant for a cooperative firm. For a cooperative we can classify variable inputs into three sets:

Set M: purchased from member patrons, e.g., corn or wheat
Set C: purchased from regional or interregional cooperatives, e.g., petroleum products
Set A: all others

The relevant total variable short-run cost function is

$$\sum_{i \varepsilon A} w_i x_i + \sum_{i \varepsilon M} w_i (1 + r_i) \, x_i + \sum_{i \varepsilon C} w_i (1 - r_i) x_i \tag{2}$$

For $i \varepsilon M$, r_i is the patronage refund paid to member patrons and for $i \varepsilon C$, r_i is the patronage refund received by the cooperative. For testing technical efficiency of use of inputs in set A, expression (1) is appropriate. But it is not appropriate for any other tests. For example if $i \varepsilon M$ and $j \varepsilon C$, the proper test equation is

$$w_i (1 + r_i) / w_j (1 - r_j) = f_i / f_j,$$

and if $i \Sigma M$ and $j \Sigma A$, the proper test equation is

$$w_i (1 + r_i) / w_j = f_i / f_j.$$

Even if we agree that minimization of expression (2) is an appropriate starting point, we find that equation (1)—which tests technical efficiency of a proprietary firm—is an inadequate test for cooperative firms. But I do not agree that (2) is an appropriate starting point. Look at $\sum w_i (1 + r_i) \, x_i$ for $i \Sigma M$. Is it desirable to minimize this? Granted it is a cost to the cooperative, but it is income to the patron–member–owners of the cooperative. I cannot accept that minimizing income to the patron–owner–members is a reasonable or desirable goal for cooperatives; it certainly violates the law and spirit of cooperation, and members would properly be angry with a cooperative that minimized their incomes in selecting the combination of variable inputs to use to produce each level of output.

In measuring a cooperative's revenues for use in testing efficiency, an analogous problem arises. We simply write a proprietary firm's gross revenue as $\sum p_i q_i$. Now let M, C, and A represent sets of items sold by a cooperative to members, regional or interregional cooperatives, and all others. We now have gross revenues for a cooperative of

$$\sum_{i \varepsilon A} p_i q_i + \sum_{i \varepsilon M} p_i (1 - r_i) \, q_i + \sum_{i \varepsilon C} p_i (1 + r_i) \, q_i.$$

$\sum p_i(1 - r_i)q_i$ equals revenue to the cooperative but costs to members. Members would reject the philosophy that their cooperative should be trying to maximize their costs.

As G. K. Chesterton once observed, "The essence of snobbery is the use of irrelevant criteria." Under this definition, economists' use of (1) to test efficiency of cooperatives is professional snobbery.

There is another issue that must be addressed. These measures of short-run efficiency are conditional upon a firm's previous choice of production-function i.e., upon its previous choice of types and sizes of machines and plant layout and size. In making these long-run decisions a firm must consider the short-run cost function for each choice. Because a cooperative has a different measure of short-run variable costs than does a proprietary firm, values of f_i and f_j computed from proprietary firms' production function are not even correct measures of f_i and f_j for the cooperatives' production function.

Our existing tests measure short-run static efficiency— efficiency in using fixed facilities put in place previously. Testing long-run efficiency is probably nearly impossible. It requires us to evaluate decisions made in the past by using only the information and methods available at the time the decisions were made.

One fundamental question is whether the published studies concern efficiency of the system or efficiency of operation of the existing system. The two issues are greatly different. The second compares, at least implicitly, various ways of operating the existing system. The first compares alternative systems, e.g., systems with different legal definitions of property rights; systems with varying proportions of cooperative, proprietary, and public ownership; or systems with varying amounts of public and private information. With rare exceptions, the studies of the U.S. grain marketing system have dealt only with the second issue: operation of the existing system.

Before becoming too deeply immersed in efficiency evaluations of the grain handling industry, we should recall that U.S. citizens objected to concentrations of economic power long before economists showed that monopoly is "less efficient" than competition, and that numerous citizens who have never been exposed to this conclusion of economists distrust monopoly. Their distrust is social, political if you wish. Sometimes a foreigner can see us more clearly than we see ourselves. A British author wrote in *The Economist* that:

[the basic inadequacy of the British government's whole attitude toward monopolization] arises perhaps from the mistaken assumption—which economists are guilty of encouraging —that political objections to monopoly can be altogether based on evidential grounds, economic or technical. The United States—which is often accused in Europe of having an exaggerated animus against monopoly, because it has a policy that quite often works—seldom falls into this trap. Its legal prejudice, *per se,* against anything calculated to restrain competition, is avowedly based, in the last resort, on social and even moral grounds; the economic efficiency that it believes competition generally promotes is the secondary justification, not the first. Primarily, American attitudes towards monopoly (public as well as private) are based upon a distrust of concentrations of economic power, irresponsible in that they are not finally accountable to the public. This does not make American antitrust legislation emotional and ineffective: it makes it at times even embarrassingly effective.

As Waugh (1954, p. 195) observed,

And actually the public may prefer to keep some known inefficiencies, rather than to adopt new methods—especially if the prospective improvements in efficiency might reduce employment, decrease price competition, or lead to greater concentration of economic power.

References

Adam, Brian D. and Dale G. Anderson. 1985. "Implications of the Staggers Rail Act of 1980 for Level and Variability of Country Elevator Bid Prices." *Proceedings of the Transportation Research Forum* 26:357-363.

Anderson, Dale G., Floyd D. Gaibler, and Mary Berglund. 1976. *Economic Impact of Rail Branch-Line Abandonment: Results of a South-Central Nebraska Case Study.* University of Nebraska-Lincoln Agricultural Experiment Station. SB541.

Babcock, Michael W. and H. Wade German. 1989. "Changing Determinants of Truck-Rail Market Shares." *The Logist. and Transp. Review* 25:251-270.

Babcock, M. W., L. O. Sorenson, M. H. Chow, and K. A. Klindworth. 1985. "Impact of the Staggers Rail Act on Agriculture: A Kansas Case Study." *Proceedings of the Transportation Research Forum* 26:364-372.

Baumel, C. Phillip; Thomas P. Drinka, Dennis R. Lifferth, and John J. Miller. 1973. *An Economic Analysis of Alternative Grain Transportation Systems: A Case Study.* Department of Transportation, Federal Railroad Administration Tech. Rep. FRA-OE-73-4.

Baumel, C. Phillip, Steven Hanson, and Robert Wisner. 1987. *Impact of Railroad Contracts on Corn, Wheat, and Soybean Bids to Elevators and Farmers.* Federal Railroad Administration; U.S. Department of Transportation, Report No. DOT-FRA-RRP-87-01, Wash., DC.

Beaulieu, Jeffrey R., Robert J. Hauser, and C. Phillip Baumel. 1985. "Regional Barge Service Demand Elasticities." *Proceedings of the Transportation Research Forum* 26:377-384.

Chow, M. H. 1986. "Interrail Competition in Rail Grain Rates on the Central Plains." *Proceedings of the Transportation Research Forum* 27:164-171.

Clow, Bradley and William Wilson. 1988. "Financial and Operating Performance of Co-operative Unit-Train Shippers in North Dakota." Report No. 234, Department of Agricultural Economics, North Dakota State University, Fargo.

Cobia, David W., William W. Wilson, Steven P. Gunn, and Randal C. Coon. 1986. *Pricing Systems of Trainloading Country Elevator Cooperatives.* Agricultural Economics Report 214, Department of Agricultural Economics, North Dakota State University, Fargo ND.

Connor, John M., Richard T. Rogers, Bruce W. Marion, and Willard F. Mueller. 1985. *The Food Manufacturing Industries.* Lexington Books. Lexington, Mass.

Dahl, Reynold P. 1989. "Changes in Grain Marketing, Market Structure, and Performance in the 1980s." Staff Paper P89-32, Department of Agricultural and Applied Economics, University of Minnesota, St. Paul. Reprinted in pp. 111-141 in Fryar, Ed (ed.) *Performance of the U.S. Grain Marketing System, 1989 Proceedings of North Central Regional Project NC-186,* Minneapolis, MN, Oct. 19-20, 1989, Department of Agricultural Economics and Rural Sociology, University of Arkansas, Fayetteville.

Dahl, Reynold P. 1990. "Structural Change and Performance of Grain Marketing Cooperatives." Staff Paper P90-54, Department of Agricultural and Applied Economics, University of Minnesota, St. Paul.

French, Ben C. 1977. "The Analysis of Productive Efficiency in Agricultural Marketing: Models, Methods, and Progress." pp. 94-208 in Martin, L. R. (ed.) *A Survey of Agricultural Economics Literature, Vol. 1.* Univ. of Minnesota Press, Minneapolis.

Fuller, Stephen and C. V. Shanmugham. 1981. "Effectiveness of Competition to Limit Rail Rate Increases Under Deregulation: The Case of Wheat Exports from the Southern Plains." *So. Jour. Agric. Econ.* 13(2):11-20.

Fuller, Stephen, Larry Makus, and Merritt Taylor. 1983. "Effect of Railroad Deregulation on Export-Grain Rates." *No. Cent. Jour. Agric. Econ.* 5(1):51-64.

Fuller, S., D. Bessler, J. MacDonald, and M. Wohlgenant. 1987. "Effect of Deregulation on Export-Grain Rail Rates in the Plains and Corn Belt." *Journal of Transportation Research Forum* 28:160-167.

Fuller, Stephen W., Fred J. Ruppel, and David A. Bessler. 1990. "Effect of Contract Disclosure on Price: Railroad Grain Contracting in the Plains." *West. Jour. Agric. Econ.* 15:265-271.

Fuller, Stephen, Orlo Sorenson, Marc Johnson, and Robert Oehrtman. 1981. "Alternative Wheat Collection and Transportation Systems for the Southern U.S. Plains." *West. Jour. Agric. Econ.* 6:91-101.

Ginder, Roger G. 1985. "Competitive Strategies of Grain Originators in Local Markets During Periods of Excess Capacity and Financial Stress." Paper presented to NCR140 *Research on Cooperatives Seminar,* Kansas City, MO. April 25.

Hanson, Steven, D., C. Phillip Baumel, and Daniel Schnell. 1989. "Impact of Railroad Contracts on Grain Bids to Farmers." *Amer. Jour. Agric. Econ.* 71:638-646.

Hanson, Steven D., Stephen B. Baumhover, and C. Phillip Baumel. 1990a. "Characteristics of Grain Elevators that Contract with Railroads." *Amer. Jour. Agric. Econ.* 72:1041-1046.

Hanson, Steven D., Stephen B. Baumhover, and C. Phillip Baumel. 1990b. "The Impacts of Railroad Transportation Costs on Grain Elevator Handling Margins." *No. Cent. Jour. Agric. Econ.* 12:255-266.

Hauser, Robert J. 1986. "Competitive Forces in the U.S. Inland Grain Transport Industry: A Regional Perspective." *Logist. and Transp. Review* 22:158-183.

Hauser, Robert J., Jeffrey Beaulieu, and C. Phillip Baumel. 1984. "Implicit Values of Multiple Car Grain Loading Facilities in Iowa and Nebraska." *No. Cent. Jour. Agric. Econ.* 6:80-90.

Hauser, Robert J. and C. Phillip Baumel. 1987. "Impacts of Market and Infrastructure Changes on the U.S. Domestic Transportation System." pp. 4-19 in Economic Research Service *Transportation and Competitiveness of U.S. Agricultural Products in World Markets: Proceedings of a Research Symposium, October 1986.* Staff Report AGES870612, U.S. Department of Agriculture, Washington, D.C.

Hilger, Donald A.; Bruce A. McCarl, and J. William Uhrig. 1977. "Facilities Location. The Case of Grain Subterminals." *Amer. Jour. Agric. Econ.* 59:674-683. Reprinted as Ch. 2 in Koo and Larson.

Hoffman, Linwood, Mack Leath, and Lowell Hill. 1985. "Effects of a Flexible Rail Rate Policy upon the Storage and Transportation System for Feed Grains." *No. Cent. Jour. Agric. Econ.* 7(1):79-93.

Holdren, Bob R. 1968. *The Structure of a Retail Market and the Market Behavior of Retail Units.* Iowa State University Press. Ames.

Johnston, J. 1960. *Statistical Cost Analysis*, New York: McGraw-Hill.

Koo, Won W. and Donald W. Larson. 1985. *Transportation Models for Agricultural Products.* Westview Press, Boulder, CO.

Koo, Won W., Stanley R. Thompson, and Donald W. Larson. 1985. "Alternative Transportation Rate and Cost Structures: A Linear Programming Model." Ch. 7 in Koo and Larson.

Ladd, George W. 1974. "A Model of a Bargaining Cooperative." *Amer. Jour. Agric. Econ.* 56:509-519.

Ladd, George W. 1983. "Value Judgments and Efficiency in Publicly Supported Research." *So. Jour. Agric. Econ.* 15:1-7.

Ladd, George W. 1988. "Costs and Goals of the Multiproduct Firm." *Managerial and Decision Economics.* 9(1988):279-81.

Ladd, George W. and Dennis R. Lifferth. 1975. "An Analysis of Alternative Grain Distribution Systems." *Amer. Jour. Agric. Econ.* 57:(420-430). Reprinted as Ch. 3 in Koo and Larson.

Lang, Mahlon G. 1980. "Economic Efficiency and Policy Comparisons." *Amer. Jour. Agric. Econ.* 62:772-777.

Larson, Donald W. and Michael D. Kane. 1979. "Effects of Rail Abandonment on Grain Marketing and Transportation Costs in Central and Southwestern Ohio." *No. Cent. Jour. Agric. Econ.* 1:105-113.

MacDonald, James M. 1987. "Competition and Rail Rates for the Shipment of Corn, Soybeans, and Wheat." *Rand Jour. Econ.* 18:151-163.

MacDonald, James M. 1987. "Developments in Grain Rail Rates and services Since Deregulation." pp. 28-38 in Economic Research Service, 1987. *Transportation and Competi-*

tiveness of U.S. Agricultural Products in World Markets: Proceedings of a Research Symposium, October 1986. U.S. Department of Agriculture, Staff Report No. AGES870612. Washington, DC.

McConnon, James C., Jr. 1989. *Methods for Assessing the Likelihood of Country Grain Elevator Failure in the United States.* Unpub. Ph.D. dissertation; Iowa State University, Ames.

Marion, Bruce W. and NC 117 Committee. 1986. *The Organization and Performance of the U.S. Food System.* Lexington Books, Lexington, Mass.

Miller, John J., C. Phillip Baumel, and Thomas A. Narigon. 1981. "Railroad Rate Deregulation: Effects on Corn and Soybean Shipments." pp. 341-364 *National Conference on Grain Marketing Patterns,* South. Coop. Series Bul. 307, No. Cent. Coop. Series Bul. 299, TVA Circular 2-173.

Miller, John J., C. Phillip Baumel, and Thomas P. Drinka. 1977. "Impact of Rail Abandonment Upon Grain Elevator and Rural Community Performance Measures." *Amer. Jour. Agric. Econ.* 59:745-49.

Moore, Frederick T. 1959. "Economies of Scale: Some Statistical Evidence," *Quart. Jour. Econ._*73, 232-249.

Musgrave, Richard. 1959. *The Theory of Public Finance: A Study in Public Economy.* McGraw Hill, New York.

Musgrave, Richard. 1986. *Public Finance in a Democratic Society.* Wheatsheaf Books, Brighton, Sussex.

Paul, Allen B. 1964. "Capital, Finance, and Market Structure: Two Approaches." pp. 19-43 in Farris, Paul L. (ed.). 1964. *Market Structure Research.* Iowa State Univ. Press, Ames.

Pazner, Elisha. 1972. "Merit Wants and the Theory of Taxation," *Public Finance.* 27: 460-472.

Royer, Jeffrey S. 1978. *A General Nonlinear Programming Model of a Producers' Cooperative Association in the Short Run.* Unpub. Ph.D. dissertation, Iowa State Univ., Ames.

Rudel, R. K. and C. E. Lamberton. 1976. *A Pilot Study to Investigate Efficient Grain Transportation and Marketing Systems for South Dakota.* Department of Transportation, Office of the Secretary. Rep. DOT-0S-50229.

Salomone, Daniel, David E. Moser, and Joseph C. Headley. 1977. *Economic Impact of Alternative Grain Transportation Systems, A Northwest Missouri Case Study.* University of Missouri Agricultural Experiment Station. Res. Bul. 1019.

Samuelson, Paul A. 1947. *Foundations of Economic Analysis,* Harvard Univ. Press, Cambridge, Mass.

Schmiesing, Brian H., Steven C. Blank, and Steven P. Gunn. 1985. "The Influence of Technological Change on Grain Elevator Pricing Efficiency." *N. Cent. Jour. Agric. Econ.* 7(1):95-104.

Solomon, Seyoum; Donald W. Larson, and Francis E. Walker. 1981. *Rail Line Abandonment: Impact on Grain Marketing and Transportation Costs in Western Ohio.* Ohio Agricultural Research and Development Center, Wooster, Res. Bul. 1131.

Sorenson, L. Orlo and Stephen W. Fuller. 1982. *An Economic Analysis of the Use of Unit Trains for Exporting Wheat from Kansas, Oklahoma, and Texas.* Kansas State University Agricultural Economics Research Report No. 82-856-D.

Thompson, Sarahelen R. and Stanley M. Dziura, Jr. 1987. "Factors Associated with Merchandising Margins at Illinois Grain Elevators, 1982-83: A Performance Analysis." *N. Cent. Jour. Agric. Econ.* 9:113-121.

Thompson, S. R., R. J. Hauser, and B. A. Coughlin. 1988. "The Competitiveness of Rail Rates for Export-Bound Grain," pp. 175-199 in Ed Fryar, (ed.), 1988, *Performance of the U.S. Grain Marketing System, 1988 Proceedings of North Central Regional Project NC-186,* Minneapolis, MN. October 20-21, Department of Agricultural Economics and Rural Sociology, University of Arkansas, Fayetteville.

VanSickle, John and George W. Ladd. 1983. "A Model of Cooperative Finance," *Amer. Jour. Agric. Econ.* 65:273-281.

VanSickle, John and George W. Ladd. 1986. "A Model of Cooperative Finance: Reply." *Amer. Jour. Agric. Econ.* 68:173-176.

Waugh, Frederick V. (ed.). 1954. *Readings on Agricultural Marketing.* Ames: Iowa State Univ. Press.

Williams, Sheldon W.; David A. Vose, Charles E. French, Hugh L. Cook, and Alden C. Manchester. 1970. *Organization and Competition in the Midwest Dairy Industries.* Iowa State Univ. Press, Ames, Iowa.

Wilson, William W., Wesley W. Wilson, and Won W. Koo. 1988. "Modal Competition in Grain Transport." *Jour. of Transp. Econ. and Policy.* 22:319-337.

The Economist. 1962. "No Policy for Mergers." February 3, pp. 393-394.

5

Structural Change in the Grain Marketing Industry

Reynold P. Dahl

Grain Marketing Channels

The grain marketing system begins at the local level with the country elevator. Country elevators have traditionally performed three important economic functions: (1) grain assembly, (2) grain storage, and (3) merchandising farm supplies and services. In some areas, large producers by-pass the country elevator and sell directly to subterminal elevators, terminal elevators, or grain processors, but the country elevator is still the primary link in the marketing system for most U.S. grain producers (Figure 5.1). Country elevators have become larger in recent years as local grain marketing cooperatives (farmers' elevators) have merged to achieve lower costs associated with larger volumes. Other smaller country elevators have gone out of business. Fewer country elevators are needed today with improvements in transportation. The grain transportation system consists of trucks, railroads, and barges on inland waterways. Changes in transportation costs as impacted by intermodal competition

Figure 5.1 • Grain Marketing, DIstribution Channels, and Modes of Transportation

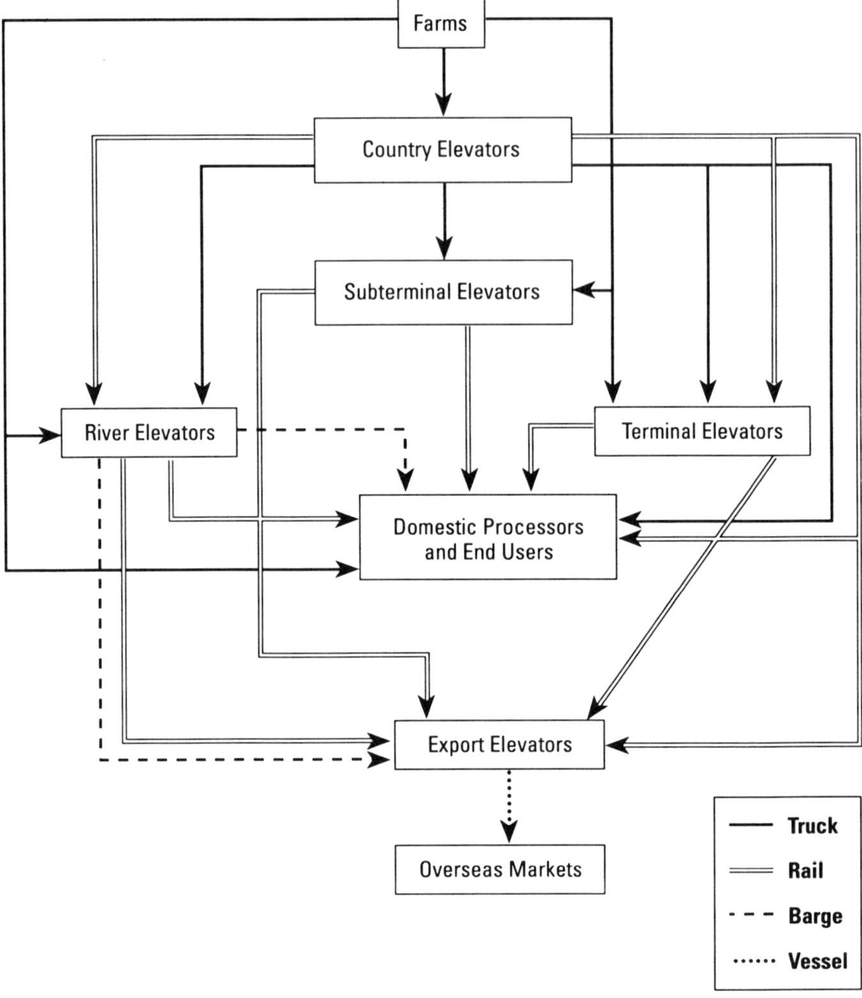

and railroad deregulation over the past two decades have been important factors inducing structural changes in the U.S. grain marketing system.

Unit Train Rates Change Structure of Country Elevators

Country elevators were first organized to perform the grain assembly function. They bought grain from surrounding farms and assembled it in quantities large enough to ship to terminal markets in single rail cars. Railroads remained the dominant mode of grain transportation until trucks came into heavy usage after World War II when a series of rail rate increases and the development of the interstate highway system made trucks highly competitive, particularly on short hauls. Country elevators began shipping large quantities of grain to terminal markets, particularly to river elevators, by truck. Grain transport by river barge also came into heavy usage at this time. Truck and barge transportation of grain dovetailed well together. Both took sizeable volumes of grain business away from the railroads.

The railroads responded to increased truck–barge competition by offering special multi-car (unit train) rates on shipments of 25, 50, 75 or more cars. Unit train rates spread rapidly as railroads were given more flexibility in rate-making under deregulation. Unit train rates were considerably lower than single car rates and provided a powerful incentive for country elevators to modernize their load-out facilities to take advantage of these lower rates. Unit train rail rates also stimulated the investment in new subterminal elevators in the country specifically designed to receive grain from other elevators, and sometimes directly from farmers, and ship it out in unit trains. Unit train shipping country elevators and new subterminal elevators spread rapidly in the 1970s. Investments were facilitated by record earnings from grain merchandising during this period providing equity capital for improvements.

A North Dakota study reported that by 1984 there were 544 unit train rail loading facilities, over half of which were farmer-owned cooperatives, in the four-state area of Iowa, Minnesota, Nebraska, and North Dakota. This represented considerable excess grain loading capacity in all of these states, particularly in Iowa, which had 5.83 bushels of unit train loading capacity for every bushel of grain shipped out of the state by rail or truck (Cobia, Wilson, Gunn, and Coon, 1986).

Excess capacity in unit train shipping facilities at the country end of the marketing system squeezed grain merchandising margins. A study by Ginder (1985) concluded that 20 percent of the local grain marketing cooperatives in the Eighth Farm Credit District were in a financially stressed condition in late 1984. He cautioned that if these firms were forced to liquidate, the market for plant and equipment in grain origination would be depressed.

Reduced income from grain merchandising associated with excess capacity and reduced grain exports in the 1980s was mitigated to some extent by increased storage income as carryover stocks accumulated under government programs. However, the precipitous drop in grain stocks as a result of the 1988 drought resulted in reduced storage income. Excess capacity in both grain storage and merchandising continues to be a problem at the country end of the marketing system.

Cash Trade at Grain Exchanges and Terminals Declines

Grain exchanges in terminal markets such as Chicago, Kansas City, and Minneapolis have played an important role in the development of an efficient grain marketing system in the United States. They brought together buyers and sellers for grain trading in a central marketplace. Open and competitive trading improved price discovery mechanisms and market information. This increased competition and broadened the market for the farmers' grain. Futures trading evolved out of cash grain trading at grain exchanges and its importance has increased over the years.

Grain commission merchants played a central role in the marketing of cash grain at grain exchanges for many years. Country elevators would consign single rail cars of grain with a sample to a commission firm that would display the sample on the trading floor and sell the grain at the highest possible price to a terminal elevator operator, processor, exporter, or other buyer. Commission firms also performed a variety of other services to the country elevator such as financing, hedging, and handling details of transportation in return for their fee. But buying and selling grain on a sample basis by commission firms has largely been replaced by forward "to arrive" cash contracts between country elevators and grain merchants where price, grade, premiums, and discounts for quality, are agreed to in the contract. The consignment method of marketing grain at grain exchanges has virtually disappeared except in a few grains such as malting barley and durum wheat where the grades are only partial indicators of grain quality. Survivors have changed their operations to become grain merchants assuming title to the grain they handle. As the marketing of cash grain by sample diminished, cash grain trade at smaller exchanges such as Duluth, St. Louis, Omaha, and Toledo declined even more sharply than at the primary futures exchanges at Chicago, Kansas City, and Minneapolis.

Today most cash grain is traded by telephone. Merchants and processors telephone bid prices each day to country elevators, usually for forward delivery. Forward selling enables country elevators to fix the price as they purchase grain from farmers and have time to schedule load-out and shipping without assuming a price risk.

Changes in transportation technology and costs accelerated the decline of cash grain trade at grain exchanges in terminal markets following World War II. First, the increased volume of grain shipped by truck by-passed terminal rail markets and was not traded at grain exchanges at all. Grain was trucked directly to processors or to river elevators for shipment on interior waterways. Second, new multi-car rates offered by the railroads to compete with increased truck–barge competition were point-to-point rates that did not include the transit privilege. Transit was an integral part of the railroad rate structure under which grain could be stopped at intermediate points between origin and final destination for inspection, storage, or processing without additional charge. The thru rate applied under transit billing. As more multi-car rates were offered by the railroads, the transit privilege was eroded and virtually eliminated. The demise of the transit privilege and deregulation of the railroads as authorized by the Staggers Act of 1980 sharply reduced the flow of grain from country points to grain exchanges in terminal markets for resale.

Decentralization of Cash Grain Trade

Most grain now moves directly from gathering points in the country to domestic users such as flour mills or to export elevators without moving through a terminal market such as Minneapolis, Kansas City, or Chicago for resale. Grain merchants are still located at grain exchanges in these markets, but trading in individual cars, or unit trains, is most likely to occur near origin points in the country rather than by sample on the grain exchange floor ("Changing Face of Breadstuffs," 1983).

In addition to diminishing the role of grain exchanges in the marketing of cash grain, railroad deregulation has diminished the role of terminal elevators at these markets, particularly terminal elevators built many years ago to handle rail grain. Many of these elevators are now obsolete for grain merchandising and are suitable only for long-term storage, primarily of government-owned grain.

Cash grain marketing has become more decentralized with subterminal elevators, located in the country and shipping grain in unit trains, taking over the functions formerly held by many older terminal elevators. Subterminals are also likely to replace many country elevators, which will continue to decline in number. One analyst projects that country elevators that are still operating 20 years from now will be subterminal elevators ("Grain Terminals Must Adapt to New Role," 1984).

Decentralization of cash grain marketing also means that terminal cash grain price quotations are not as representative of true cash grain prices as in years past because they are based on a smaller volume of trade. Cash grain prices are now determined more at export locations than at terminal markets. Futures prices have become even more important as a "basis" for pricing cash grain in a marketing system that has become more decentralized.

Futures Trading in Grain Reaches New High

Price volatility increased during the grain export boom of the 1970s as grain prices rose and the U.S. government was able to dispose of stocks that had been accumulated in the post-World War II period under price support operations. This increased hedging needs, which pushed the volume of futures trading in grain to a record high of 39.5 million futures contracts in 1980 (Table 5.1). Marketing decisions in volatile grain markets emerged as new and complex problems for farmers as well as marketing firms. Agricultural marketing economists directed more of their attention in both teaching and research to futures markets, hedging, and price risk management.

Futures trading in grain and products varies positively with price variability and inversely with government price support loan activity. As grain exports declined in the early 1980s, price volatility was reduced and stocks accumulated under government programs increased. The volume of futures trading in grain declined from its record level of 39.5 million contracts in 1980 to 26.8 million contracts in 1987; reflecting reduced hedging needs associated with lower price variability and the accumulation of grain stocks to record levels under government programs. But futures trading in grain rebounded in 1988 to reach a new record of 40.9 million contracts as prices and price volatility increased with the drought and the precipitous draw-down in grain stocks. The volume of futures trading has continued at the high level of 39.5 million futures contracts in 1994 (Table 5.1).

Issues in Futures Market Performance

The declining importance of terminal markets in cash grain trade is particularly relevant to recent questions about the adequacy of deliverable

Table 5.1 • Futures Contracts Traded on U.S. Grain Futures Markets, by Commodity, Selected Years

Exchange and Commodity	Contract Unit	Contracts (Thousands)				
		1973	1980	1987	1988	1994
Chicago Board of Trade						
Wheat	5,000 bu.	1,567	5,428	1,929	3,378	3,621
Corn	5,000 bu.	4,075	11,947	7,253	11,106	11,530
Oats	5,000 bu.	183	321	291	355	493
Soybeans	5,000 bu.	2,743	11,768	7,379	12,497	10,749
Soybean oil	60,000 lb.	1,763	3,168	3,912	4,896	5,063
Soybean meal	100 tons	660	3,219	3,798	5,313	4,594
Rice		0	0	0	0	17
Total		10,991	35,851	24,562	37,545	36,067
Kansas City Board of Trade						
Wheat	5,000 bu.	346	1,298	971	1,339	1,502
Minneapolis Grain Exchange						
Spring wheat	5,000 bu.	172	334	311	424	737
White wheat	5,000 bu.	0	0	1	*	27
Total		172	334	312	424	764
Mid-America Commodity Exchange						
Wheat	1,000 bu.	75	551	190	294	102
Corn	1,000 bu.	103	441	312	429	233
Oats	1,000 bu.	9	2	7	13	5
Soybeans	1,000 bu.	56	1,053	418	864	798
Soybean meal	20 tons	0	0	3	9	4
Rice		0	0	0	0	68
Total		243	2,047	930	1,609	1,210
Total all markets		11,752	39,530	26,781	40,919	39,543

*Less than 1,000 contracts.
Source: Futures Industry Association.

stocks against Chicago Board of Trade wheat, corn, and soybean futures contracts. These questions followed the Chicago Board of Trade emergency action taken in July 1989 that ordered the liquidation of the largest positions in the July 1989 soybean futures contract. This action was necessitated by a number of facts known at that time and later reported in a study by the Commodity Futures Trading Commission. First, the open interest on 10 July, 1989, in the July future was unusually large, over 40 million bushels, and deliverable stocks were very low, less than 13 million bushels, due in large part to the drought of 1988. Second, both the long side of the July contract and the supply of deliverable soybeans were held, in large proportions, by the same trader. Third, the July futures contract was priced more than 40 cents higher than the August future on 10 July. The Chicago Board of Trade took the emergency action to assure an orderly liquidation of the July soybean contract to prevent severely distorted July soybean futures prices that could have resulted in contract defaults (Hineman, 1989). The impact of this emergency action on the soybean market precipitated widespread controversy.

The National Grain and Feed Association commissioned a study by Peck and Williams of Stanford University entitled "An Evaluation of the Performance of the Chicago Board of Trade Wheat, Corn, and Soybean Futures Contracts During the Delivery Periods from 1964–65 Through 1988–89." The results of this study are significant and worthy of emphasis. First, deliveries against CBOT grain futures contracts are more important than generally believed: "CBOT wheat, corn, and soybean markets have delivery on the order of 10 percent or 20 percent of the peak open interest. Moreover, of these positions still outstanding on the day just before the delivery period, as many as 50 percent are satisfied through actual delivery." Second, there is a significant concentration of positions of the four largest traders with long and short positions at the start of and during the delivery period. This concentration along with the decline in deliverable stocks has reduced the price spread (carrying charge) between the expiring contract month and the next contract month. Third, "basis convergence in Chicago has deteriorated from the 1960s to the 1980s." Deliveries on the three CBOT contracts have been increasing as a percentage of deliverable stocks. This was considered evidence that stocks were too low. The study attributed the low level of deliverable stocks to the decline in terminal markets in cash grain trade. It rejected two proposed solutions to inadequate deliverable stocks, namely, more terminal delivery points and cash settlement. The study suggested a re-evaluation of the delivery of grain in store and allowing for barge delivery or on-track delivery at Gulf export terminals.

An important characteristic of a good futures contract is that its terms, including delivery provisions, reflect the realities of commercial trade. This enables futures contracts to serve as temporary substitutes for later cash contracts on other terms. The realities of commercial trade in wheat, corn, and soybeans do not reflect as much movement through Chicago as was true in years past. Today, the largest share of export movements of wheat, corn, and soybeans is by rail and water to Gulf and Pacific–Northwest ports (Hill and Timmerman). Domestic processors also obtain supplies directly in the country and not through terminal markets such as Chicago.

Changes in Delivery on Chicago Board of Trade Grain Futures

The Chicago Board of Trade announced the following changes in the delivery provisions of its wheat, corn, and soybean futures contracts in the spring of 1992: (1) the addition of St. Louis as a delivery location for soybeans and wheat (St. Louis was already a delivery point for corn); (2) delivery price differentials at St. Louis were set at premiums to Chicago; (3) increases in minimum performance standards by delivery warehouses (load-out, load-in rates for rail, barge, and vessels, responsibility for insuring grain, etc.); and (4) limits to storage rates charged by delivery warehouses. ("Ag Contacts Undergo Changes in Delivery," 1992.)

U.S. Largest Multiple Facility Grain Firms

The 10 largest U.S. grain companies operated 914 grain facilities with aggregate storage capacity of 1,749 million bushels as listed in the *1995 North American Grain and Milling Directory* (Table 5.2). The facilities included 42 port, 139 river, 140 terminal, 147 subterminal, and 446 country elevators. The distinction between the latter two types of facilities is often difficult and numbers can vary with interpretation. Some country elevators in Table 5.2 would undoubtedly be classified as subterminals if the latter is defined as an elevator located in the grain-producing area that receives grain from other elevators, and sometimes directly from farmers, and has the capability of loading and shipping the grain in unit trains. Such elevators have increased in both number and importance in response to special unit train rates offered by the railroads. This trend will likely continue as more grain moves directly from country gathering points to ports or to domestic processors without moving through termi-

Table 5.2. Ten Largest U.S. Multiple Facility Grain Companies by Number of Facilities and Capacity, 1995.

Company	Number of Grain Storage Facilities					Total Number	Total Storage Capacity (mil. bu.)
	Port	River	Terminal	Sub-terminal	Country		
1. Cargill, Inc. (total)	14	21	9	16	160	220	398
Cargill, Inc. (grain div.)	(14)	(19)	(7)	(0)	(160)	(200)	(360)
Cargill, Inc. (flour milling div.)	(0)	(2)	(2)	(16)	(0)	(20)	(38)
2. Archer Daniels Midland Co. (total)	3	25	54	27	107	216	379
ADM Grain Co.	(3)	(25)	(18)	(27)	(107)	(180)	(327)
ADM Milling Co.	(0)	(0)	(36)	(0)	(0)	(36)	(52)
3. Con Agra, Inc. (total)	6	16	16	50	28	116	191
Peavey Co. (subsidiary)	(5)	(13)	(8)	(32)	(28)	(86)	(151)
Grain Processing Co.	(1)	(3)	(8)	(18)	(0)	(30)	(40)
4. Continental Grain Co.	8	26	18	15	13	80	180
5. Bunge Corp.	3	41	5	6	*	55*	173
6. Farmland Grain Division (Division of Farmland Industries, Inc.)	1	1	10	0	11	23	111
7. Riceland Foods, Inc.	0	2	3	29	0	34	95
8. Country Mark Cooperative, Inc.	3	2	12	3	0	20	87
9. Harvest States Cooperatives	2	5	2	0	100	109	69
10. General Mills, Inc.	2	0	11	1	27	41	66
Ten Largest Total	42	139	140	147	446	914	1,749

Source: *1995 North American Grain and Milling Directory*, Sosland Publishing Co.

*Does not include country elevators.

nal markets. The data shown in Table 5.2 are not available for previous years.

The ten largest U.S. grain elevator companies in 1981 are shown in Table 5.3, but country elevators are excluded. Only data for subterminal, terminal, river, and port elevators are included in the number of elevators and storage capacity. Nevertheless, one can compare Tables 5.2 and 5.3, and note that many changes have occurred.

Cargill maintained its lead as the nation's largest grain company from 1981 to 1995. However, its lead has been challenged in recent years by the Archer Daniels Midland Co. (ADM). This firm has expanded rapidly through both internal growth and acquisitions. In fact, *Milling and Baking News* reported in October 1993 that ADM became the nation's largest grain company with the establishment of a joint venture under which ADM assumed operational control of most U.S. grain elevators owned by Louis Dreyfus Corp. Grain storage capacity of Louis Dreyfus Corp. totaled 77 million bushels, including 12 terminal elevators, 2 port facilities, 9 river facilities, 3 subterminals, and 23 country elevators.

Con Agra, Inc. ranked as third largest in 1995. Like ADM, it too expanded through acquisitions such as the Peavey Co. and the grain merchandising operations of the Pillsbury Co. The Continental Grain Co. and the Bunge Corp. ranked as the fourth and fifth largest U.S. grain companies.

Farmer-owned cooperatives, namely, the Farmland Grain Division of Farmland Industries, Inc., Riceland Foods, Inc., Country Market Cooperative, Inc., and Harvest States Cooperatives ranked sixth through ninth, respectively, with General Mills, Inc. ranking tenth.

Structural Changes in Grain Marketing Cooperatives

The Agricultural Cooperative Service (ACS) of the U.S. Department of Agriculture reported a total of 16 regional grain marketing cooperatives and three interregionals in the United States in 1981 with a total grain handle of 3 billion bushels (Thurston and Cummins, 1983). This was their zenith year that also marked the beginning of a decade in which a downsizing of these farmer-owned cooperatives would occur (Dahl, 1991).

One interregional grain marketing cooperative (The Farmers Export Company), a federation of regional grain marketing cooperatives organized to market farmers' grain for export, was liquidated in 1985 through

Table 5.3 Ten Largest U.S. Grain Elevator Companies, 1981.

Company	Number of Elevators[1]	Total Storage Capacity (million bu.)
1. Cargill, Inc.	21	148
2. Far-Mar-Co, Inc.	17	122
3. Continental Grain Co.	39	110
4. Union Equity Co-op Exchange	3	67
5. The Pillsbury Co.	44	54
6. Central Soya Co.	9	51
7. Bunge Corp.	51	47
8. The Anderson's	7	43
9. Lincoln Grain, Inc.	3	39
10. Indian Grain Division	12	39
Ten Largest Total	206	720

Source: *Milling and Baking News,* Sosland Publishing Co. (13 October 1981).
[1] Port, river, terminal and subterminal elevators. Does not include country elevators.

the sale of its remaining assets to the Archer Daniels Midland Company (ADM). An interregional river barge transportation cooperative (Agri-Trans Corporation) was also re-structured in 1986 when it was merged with the American River Transportation Company (ARTCO), a subsidiary of ADM.

Two regional grain marketing cooperatives, the Producers Grain Corporation of Amarillo, Texas, and Far-Mar-Co., of Hutchinson, Kansas, closed their grain marketing operations in the 1980s. Earlier, Far-Mar-Co. had merged with Farmland Industries of Kansas City, becoming a subsidiary of this regional farm supply cooperative. A sizeable share of Far-Mar-Co.'s grain marketing assets were later sold to the Union Equity Co-op Exchange of Enid, Oklahoma.

Two well-known regional cooperatives in the cornbelt, GROWMARK, of Bloomington, Illinois and AGRI Industries of Des Moines, Iowa, transferred their grain marketing operations to joint ventures with major multinational corporations. GROWMARK transferred ownership of its seven river terminals to a new ADM subsidiary called ADM/GROWMARK in exchange for ADM common stock (GROWMARK and ADM Announce Plans for Joint Venture, 1985).

AGRI Industries also formed a joint venture with Cargill, Inc. called AGRI Grain Marketing. With the integration of AGRI's grain merchandising and related functions into the new joint venture, AGRI Industries became a holding company "functioning as a cooperative enterprise in supporting member services and other cooperative programs" (Coonrod, 1986).

Two mergers of regional grain marketing cooperatives also occurred during the 1980s. The Grain Terminal Association, St. Paul, Minn, and North Pacific Grain Growers, Inc., Portland, Oreg, merged to form Harvest States Cooperatives on 1 June 1983. The new cooperative, headquartered in St. Paul, Minn., became the nation's largest grain marketing cooperative. Harvest States has grain export facilities on the Great Lakes at Duluth Superior and the Pacific Northwest at Kalama, Wash. In 1994 Harvest States Cooperatives also purchased an export terminal at Myrtle Grove, La. giving it access to grain export at the Mississippi Gulf. Harvest States Cooperatives serves farmers in the upper Midwest, Pacific Northwest, and adjoining areas.

Ohio Farmers Grain and Supply Association merged with Landmark, Inc. to become Countrymark, Inc. in 1985. Countrymark then purchased the assets of Agra Land, the cooperative that emerged in 1983 from the Chapter 11 bankruptcy of Michigan Farm Bureau Services. Mid-States Terminals, Inc. (the fourteenth largest multiple facility grain firm in the United States.) then became a wholly owned grain subsidiary of Countrymark, Inc. (Benschneider, 1987).

Countrymark, Inc. and the Indiana Farm Bureau Cooperative Association (IFBCA), both federated agricultural supply and grain marketing cooperatives serving farmers in Ohio, Michigan, and Indiana, merged to form Countrymark Cooperative, Inc. in 1995 (Duffy, 1991, pp. 20-21).

Farmland Industries, Inc. re-entered the grain business in 1992 through the purchase of Union Equity Co-op Exchange to form the Farmland Grain Division. Later it purchased Tradigrain, a Swiss-based company considered a major player in global grain trading ("Farmland Expects Added Value from Return to Grain Trading," 1994).

The organization structure of U.S. grain marketing cooperatives is diagrammed in Figure 5.2. The total number of grain marketing cooperatives, most of which are local cooperatives, in the United States declined from 1,623 in 1985 to 1,193 in 1993 (*Farmer Cooperative Statistics*,

1993). Regional cooperatives are federations of locals. Downsizing and consolidation have left four regional cooperatives serving farmers in the nation's principal grain growing areas: (1) Countrymark Cooperative, Inc. in the eastern cornbelt, (2) Harvest States Cooperatives in the upper Midwest and Pacific Northwest, (3) Farmland Grain Division (Division of Farmland Industries, Inc.) in the Southern Plains, and (4) Riceland Foods, Inc. in the south central region of the United States.

A new interregional grain processing cooperative, Ag Processing, Inc. (AGP) was established in 1983 through the consolidation of the soybean processing operations of Boone Valley Cooperative Processing Association, Land O'Lakes, Inc. and Farmland Industries, Inc. On 31 August 1991, AGP had 11 regional cooperative members and 381 active local cooperative patrons. In 1991, AGP purchased a grain terminal and formed a subsidiary, AGP Grain Cooperative. It also acquired the North American feed manufacturing facilities, ag service centers, and grain operations of International Multifoods through a joint venture with ADM (Ingalsbe, 1992, p. 9).

Joint ventures between farmer-owned cooperatives and investor-oriented firms (IOF's) are new structural innovations in grain marketing. As explained above, two regional cooperatives in the cornbelt, GROWMARK, Inc. and AGRI Industries, Inc., transferred their grain marketing operations to joint ventures with major multi-national corporations. Other joint ventures with IOF's have been organized in the 1990s. The expansion of joint ventures in the United States grain marketing system is analyzed later in this paper.

Changes in U.S. Grain Export Market Structure

A major study by Conklin published by the U.S. General Accounting Office in 1982 categorized the market structure of the U.S. grain export system into four groups: (1) major multi-national corporations, other than Japanese, (2) Japanese-owned or affiliated firms, (3) farmer-owned cooperatives, and (4) all other exporting firms ("Market Structure and Pricing Efficiency of U.S. Grain Export System," 1982).

Major multi-national corporations are large firms which operate globally and handle much of the grain that is bought and sold in the world today. The five largest multi-nationals in 1980–81 were Cargill, Inc.; Continental Grain Company; Bunge Corp.; Louis Dreyfus Corp.; and Garnac Grain Co., Inc.

Japanese firms are likewise multi-national in nature and some are large. Trading firms such as Marubeni, Mitsui, Mitsubishi, and C-Itoh play an important role in exporting U.S. grain to Japan and other countries. Some of these firms have also acquired U.S. facilities including country elevators, subterminals, terminals, and port elevators.

The Japanese National Federation of Agricultural Cooperative Associations (Zen-Noh) also established Zen-Noh Grain Corp., a U.S. subsidiary, which constructed a modern grain export terminal at Covenant, Louisiana, in 1982. Its purpose is to purchase corn, soybeans, and milo from American farmers and ship these grains to Japan ("Zen-Noh's U.S. Elevator," 1983).

A grain export firm is typically defined as a firm that sells grain directly to a foreign buyer. It does not necessarily have to load the grain on an ocean-going vessel, because this is sometimes done by another company. The *Grain Guide 1988* listed 61 U.S. grain exporting companies. Included are the large multinationals, referred to above, and other U.S. corporations, cooperative and non-cooperative, well-known in the grain business. A sizeable number of Japanese firms are also listed. Other firms listed are not widely known in the U.S. grain business and provide evidence that small as well as large firms can participate in the U.S. grain export business. This is contrary to the popular view that heavy capital requirements are barriers to entry in grain exporting.

Farmer-owned cooperatives downsized grain operations in the mid-1980s, but have expanded direct grain exporting in the 1990s. Harvest States Cooperatives purchased an export terminal in Myrtle Grove, La. in 1994 which is jointly operated with another Gulf export terminal owned by the Peavey Co. in a joint venture (HSPV). Harvest States Cooperatives reported handling over 1 billion bushels of grain in fiscal 1995, of which 50 percent went for direct export sales (*1995 Harvest States Annual Report*).

The Farmland Grain Division of Farmland Industries, Inc. has also expanded its direct grain exporting through the purchase of Tradigrain, a Swiss-based company considered a major player in global grain trading.

The GAO study of 1982 concluded that the changing market structure of the U.S. grain export industry was inconsistent with the static make-up one would find in a monopolized industry. New firms, both large and small, have entered the industry, and others have exited, according to the study. The composition and market shares of firms in the industry have also changed significantly during this period, and these structural changes indicated competitive forces at work in the U.S. grain export system.

The above conclusions would likely apply to the U.S. grain export system today as they did a decade ago. In fact, excess capacity in the system today has squeezed marketing margins and intensified competition. The surplus in grain exporting capacity has been estimated at close to 50 percent ("Facing Up to Terrible Dilemma in Grain Trade," 1991). Another study reported that 16 of 56 export grain elevators in the United States are closed in 1991, and seven are for sale (Kimle and Hayenga, 1991). Additional evidence of excess in grain export capacity is formation of joint ventures between grain exporting companies.

The Day of the Joint Venture

Several years ago a novel entitled, *The Day of the Jackal* was published. It later was made into a popular movie. We are reminded of this title today when analyzing the U.S. grain business because an appropriate description might be "The Day of the Joint Venture." Scarcely a week goes by without a report of some new joint venture in grain handling, merchandising, or processing ("Partnering as Newest Mode for Grain Foods," 1992).

A list of joint ventures in grain marketing in existence in 1995 is shown in Table 5.4. No claim is made that the list is all inclusive. These

Table 5.4. Joint Ventures in Grain Marketing and Processing, 1995.

A. Joint ventures between cooperatives and investor-oriented firms (IOF's)

    1. *Intrade, Inc.*
Consortium of 11 German, French, Dutch, Canadian and U.S. cooperatives and the Archer Daniels Midland Company purchased the equity capital of A. C. Toepler International, a multi-national grain trading company. Grain export merchandizing.

    2. *ADM/GROWMARK, Inc.*
GROWMARK, a regional farm supply and grain marketing cooperative, Bloomington, IL, and Archer Daniels Midland Co. ADM and GROWMARK terminals on Illinois and Mississippi Rivers. Originate and merchandize grain.

3. *AGRI Grain Marketing*

   AGRI Industries, Inc., a regional grain marketing cooperative, West Des Moines, IA, and Cargill, Inc., Minneapolis. Merchandise grain and oilseeds by truck, barge, and rail through six river terminal elevators in Iowa and Illinois.

4. *Ag Processing, Inc./ADM*

   Ag Processing, Inc., an interregional soybean processing cooperative, Omaha, NE, and the Archer Daniels Midland Company purchased 18 feed plants, 26 retail ag centers, and 18 grain elevators from International Multifoods, Inc. Grain and feed merchandising.

5. *TEMCO (Tacoma Export Marketing Co.)*

   A joint venture between Continental Grain Co. and Harvest States Cooperatives: "It seeks to combine Continental's strengths in international marketing and the capabilities of its 3 million bushel Tacoma, WA export terminal with Harvest States' strengths in grain origination and logistical management."

6. *HSPV*

   A joint venture between Harvest States Cooperatives and the Peavey Co.; operates three river grain elevators in Iowa and two export terminals in Louisiana.

B. Joint ventures between investor-oriented firms

1. *M.F.P.*

   Con Agra Grain Companies, Omaha, and Ferruzzi Group, New York, form joint venture in late 1991 to export grain from the Gulf of Mexico. The joint venture combines and manages the Gulf grain export facilities of both companies.

2. *Conti/Bunge Export Marketing Group*

   Continental Grain Co., New York, and Bunge Corp., St. Louis, form joint venture in March 1992 to market and export grain through the Louisiana Gulf. Objective is to maximize utilization of Continental's port elevator at Westwego, LA, and Bunge's port elevator at Destrehan, LA. ("Continental Bunge in venture to export from Louisiana elevators," 1992.)

3. *Archer Daniels Midland Co. and Louis Dreyfus Corp.*

   A joint venture under which ADM assumed operational control of most of the U.S. grain elevators of the Louis F. Dreyfus Corp.

joint ventures have been announced in news releases seen by the author. They are divided into two groups: (1) joint ventures between farmer-owned cooperatives and investor-oriented firms (IOF's), and (2) joint ventures between investor-oriented firms.

Joint ventures with IOF's enable farmer-owned cooperatives to leverage their equity capital, pool risks, expand markets, and obtain advantages of vertical integration. To strengthen their position in international marketing, a consortium of 11 German, French, Dutch, Canadian, and U.S. cooperatives and the Archer Daniels Midland Company formed a joint venture called Intrade, Inc. in the early 1980s. It purchased the equity capital of A. C. Toepler International, a multi-national trading company headquartered in Hamburg, Germany. Intrade is a vehicle through which participating cooperatives along with ADM compete with the large multi-national grain trading companies in international markets.

Joint ventures also enable the participants to capture the strengths of each of the participants. When ADM/GROWMARK, Inc. was formed in 1985, the chief executive officer of GROWMARK described the advantages of the joint venture as follows: "ADM needs and wants our systems grain origination capability, and we need ADM's ability to provide equity capital, their processing capability, and worldwide marketing expertise" ("GROWMARK and ADM Announce Plans for Joint Venture," 1985).

A similar advantage was attributed to a new joint export venture between the Continental Grain Co. and Harvest States Cooperatives announced in August 1992. The new joint venture called TEMCO (Tacoma Export Marketing Co.) "seeks to combine Continental's strengths in international marketing and the capabilities of its 3 million bu. Tacoma, Washington, export terminal with Harvest States' strengths in grain origination and logistical management" ("Continental Grain, Harvest States in Export Venture," 1992).

Other economic advantages have been cited for other joint ventures. Con Agra Grain Companies and the Ferruzzi Group formed a joint venture (M.F.P.) for exporting grain from the Gulf of Mexico; it was emphasized that the agreement allows each firm to concentrate on the functions they perform best. Mr. Smith of Ferruzzi said, "Peavey (Con Agra's subsidiary) will be the principal partner in originating and transporting grain to the export point, and Ferruzzi will be the principal partner in supplying its industry with raw material that it needs as well as supplying the world market with U.S. grain" ("Con Agra Grain, Ferruzzi Group in Pact to Export Grain from Gulf," 1991).

Joint ventures may also serve as a first step in full acquisition by one of the partners. For example, a joint venture entitled ADM/TPC Milling was set up in 1992 when The Pillsbury Co., a subsidiary of Grand Metropolitan P.L.C., sold a 50 percent interest in its four remaining flour mills to ADM. This joint venture was operated by ADM, but was later dissolved when ADM purchased all of the equity from Pillsbury.

Joint ventures may also have disadvantages. It remains to be seen whether or not joint ventures between farmer-owned cooperatives and IOF's are a threat to the favorable legislative and regulatory treatment under which farmer-owned cooperatives now operate. Problems may also arise in connection with the inter-firm pricing of products, cost, and profit-sharing. Joint ventures between large firms in concentrated industries may also raise antitrust questions.

Grain Storage Capacity Increases

The first national survey of grain storage facilities in the U.S. was made in 1978. It showed aggregate farm and off-farm storage capacity at nearly 17 billion bushels made up of 10 billion bushels of storage on-farm (59 percent of the total) and 7 billion in off-farm facilities (41 percent of the total). This was equivalent to a full year and a half of grain production in the United States, which was about 12 billion bushels per year in 1978 (Table 5.5).

As grain exports declined in the 1980s, stocks accumulated despite sizeable acreage idled under federal farm programs. Grain stocks reached an all-time high of 8.4 billion bushels at the end of the 1986–87 marketing year. Most of these stocks were stored under government programs such as the farmer-owned reserve, regular price support loan, and CCC ownership.

Grain storage capacity increased in response to the stock build-up, reaching a record 22.9 billion bushels on 1 December 1988, an increase of 36 percent from 10 years earlier. The total of on-farm storage capacity of 13.3 billion bushels (58 percent of the total) and off-farm capacity of 9.6 billion bushels (42 percent of the total) reached about two years of U.S. grain production.

The reality, surprising as it might be, that nearly six out of ten bushels of U.S. grain storage capacity represents farm storage, reflecting the

Table 5.5 • Grain Storage Capacity in the U.S., On-Farm and Off-Farm, by State, 1 April 1978 and 1 December 1993.

State	On-Farm	Off-Farm (commercial) April 1, 1978 (millions bu.)	Total	On-Farm	Off-Farm (commercial) December 1, 1993 (millions bu.)	Total
Iowa	1,492	635	2,127	1,700	972	2,672
Illinois	1,154	787	1,941	1,200	1,123	2,323
Minnesota	1,192	368	1,560	1,325	490	1,815
Nebraska	833	488	1,321	1,000	672	1,672
Kansas	370	831	1,201	400	836	1,236
Texas	264	838	1,102	200	885	1,085
North Dakota	691	142	833	810	243	1,053
Indiana	507	283	790	700	350	1,050
Wisconsin	437	130	567	445	180	625
Missouri	347	210	557	410	256	666
Others	2,637	2,275	4,912	3,435	2,484	5,919
Total	9,924	6,987	16,911	11,625	8,491	20,116

Source: U.S. Department of Agriculture.

steady expansion of these facilities in recent years under farm program incentives. Farmers found it advantageous to have farm storage to participate in the regular nine-month farm price support program. The farmer-owned reserve, a three-year loan program provided by Congress in the 1977 Farm Bill, also provided a big boost to new farm storage. Finally, having their own storage gives farmers more flexibility in grain marketing.

Total grain storage capacity, both on-farm and off-farm, declined in the 1990s. This was in response to the record decline in U.S. grain stocks as a result of the drought of 1988. Ending U.S. grain stocks declined from their record level of 8.4 billion bushels in 1986–87 to 1.9 billion bushels at the end of the 1993–94 marketing year. Total U.S. grain storage capacity was 20.1 billion bushels on 1 December 1993, down about 2.8 billion bushels from the record level of 1988.

Eight states now have over one billion bushels in total grain storage capacity. Iowa ranks first in grain storage capacity with 2,672 million bushels, followed by Illinois, 2,323 million; Minnesota, 1,815 million; Nebraska, 1,672 million; Kansas, 1,236 million; Texas, 1,085 million; North Dakota, 1,053 million; and Indiana, 1,050 million (Table 5.5).

As grain stocks accumulated in the 1980s under federal farm programs, the grain trade derived more income from storage and handling grain for the government. The income from such operations offset, in part at least, declines in income associated with reduced grain exports and marketing margins. But the precipitous drop in grain stocks as a result of the 1988 drought resulted in excess grain storage capacity and reduced storage income for the grain marketing system.

Vertical Integration Expands in the Grain Industry

Excess capacity and reduced profit margins in the grain industry have made it difficult for the free-standing grain firm to survive on grain merchandising and storage income alone. Hence, most of the new investments in the industry have been in value-added grain processing operations. Several of the largest multiple facility grain merchandising firms have made sizeable investments in wheat flour milling, wet corn milling, livestock feed manufacturing, meat and poultry production, and processing. These investments have come in the form of acquisitions as well as in new plants and equipment.

The three largest multiple grain companies, ADM, Cargill, Inc., and Con Agra, Inc., are now the three largest wheat flour millers, operating about 60 percent of the U.S. flour milling capacity. ADM and Cargill, Inc. are also among the largest firms in soybean processing and wet corn milling. Con Agra, Inc. is also a major manufacturer of consumer foods. Vertical integration in the grain industry is illustrated by a major restructuring announced by Cargill, Inc. in early 1992. Of particular interest is the restructuring of its former Commodity Marketing Division into the new Cargill Grain Division (CGD). The CGD became the primary source of grain procurement for merchandising to both domestic and export markets, as well as for grain milling and oilseed processing operations of the company. In effect, the new Cargill Grain Division integrated the grain merchandising and grain processing operations of the company. It was also reported that Cargill, Inc. plans to reduce its reliance on commodities and increase its manufacturing and marketing operations in branded food products.

Cooperatives Emphasize Value-Added Grain Processing

Surplus capacity and low returns in grain merchandising have also caused cooperatives to expand into value-added grain processing businesses. Harvest States Cooperatives, the nation's largest grain marketing cooperative, is an interesting case in point. Its CEO reported at its annual meeting in November 1991 that the cooperative will continue to expand its profitable half-billion dollar processed food business to soften its reliance on grain trading. Harvest States Holsum Foods Division manufactures margarines, salad dressings, peanut butter, and other consumer foods derived from vegetable oils. This is a vertical extension of its Honeymeade Processing Division that produces and refines soybean oil and meal. The Annual Report of Harvest States reported that in fiscal 1995 Holsum acquired Saffola Quality Foods of Los Angeles. Combined with the 1992 purchase of Gregg Foods at Portland, the Saffola purchase made Holsum the second largest firm in the West Coast market in terms of margarine production capacity.

The Amber Milling Division of Harvest States Cooperatives that grinds durum wheat into semolina, the chief ingredient in pasta, has also been expanded. Several years ago it acquired a durum mill in Huron, Ohio, that was modernized and expanded from 6,000 cwt to 14,000 cwt per day. Harvest States completed building a large new flour mill at Kenosha, Wisconsin, early in fiscal 1996. Amber Milling later announced plans for two other new mills, one in Eastern Pennsylvania and the other at Houston, Texas. The new plants will enable Amber Milling to mill hard wheat for bread flour as well as semolina and durum flour. When these new mills are added to existing mills at Huron, Ohio, and Rush City, Minnesota, Amber Milling will extend its lead as the nation's largest durum miller serving the pasta industry.

Value-added activities, such as the manufacture of livestock feed and large-scale contractual hog and poultry feeding by their members, are also receiving considerable interest by many local grain marketing cooperatives in the upper Midwest. A number of new cooperatives have been formed to engage in value-added processing of agricultural commodities and others are being proposed (Campbell, 1995, pp. 10-16). Most of these new ventures are viewed as higher return alternatives to marketing low value grain. Such activities demand a new set of management skills in addition to those required for grain and farm supply merchandising.

But, they signify that more country elevators recognize that it may be difficult to survive in the longrun with the narrow profit margins that currently prevail in grain merchandising.

Summary of Major Economic Forces Causing Structural Change in the Grain Industry

Surplus capacity and low profit margins in grain merchandising have stimulated many mergers, acquisitions, and plant closures. The late 1980s and early 1990s can best be characterized as a period of consolidation and increased concentration in the grain industry.

Structural changes in the grain industry have been stimulated and hastened by transportation deregulation and unit train rail rates. Most grain now moves directly from gathering points in the country to domestic users such as flour mills or to export elevators without moving through a terminal market for re-sale. Cash grain marketing has become more decentralized with subterminal elevators located in the country taking over the functions formerly held by country elevators and rail terminals, many of which have become obsolete.

The volume of futures trading in grain and products has increased with increased hedging needs. Futures prices have become even more important as a "basis" for pricing cash grain. The delivery provisions on grain futures contracts need surveillance as reduced volumes of grain flow through terminal markets such as Chicago, Kansas City, and Minneapolis.

Joint ventures are new structural innovations in the grain industry and can be classified as: (1) joint ventures between regional grain cooperatives and investor-oriented firms (IOF's), and (2) joint ventures between IOF's. Several have been dissolved after serving as a preliminary step toward full acquisition by one of the partners. Joint ventures have advantages as well as disadvantages.

Vertical integration is proceeding rapidly in the grain industry. Most of the new investments in the industry are in value-added grain processing activities by farmer-owned cooperatives such as Harvest States Cooperatives and IOF's such as Cargill, Inc., ADM, and Con Agra, Inc. The integration of grain merchandising, exporting, and grain processing is proceeding at an accelerated pace.

References

Benschneider, Donald E. "The Creation of Countrymark, Inc.," *American Cooperation 1987*, pp. 244-48. Washington, D.C.: American Institute of Cooperation, 1987.

Campbell, Dan. "Co-op Fever is Still Sizzling Across North Dakota," *Farmer Cooperatives*, USDA/Rural Business Development Services, August 1995.

Cobia, David, William Wilson, Steven Gunn, and Richard Coon. *Pricing Systems of Trainloading Country Elevator Cooperatives*. Agri. Econ. Rep. 214, Dept. of Agri. Economics, Agri. Exp. Stat., North Dakota State University, December 1986.

Coonrod, Richard A. "Letter to All Member Companies." West Des Moines, Iowa: AGRI Industries, Inc., February 11, 1986.

Dahl, Reynold P. "Structural Change and Performance of Grain Marketing Cooperatives." *Journal of Agricultural Cooperation*, 5(1991): 66-80.

Duffey, Patrick. "Indiana, Countrymark Merging Common Cooperative Interests." *Farmer Cooperatives*, July 1991, pp. 20-21.

Ginder, Roger G. "Competitive Strategies of Grain Originators in Local Markets During Periods of Excess Capacity and Financial Stress." Paper presented at NCR-140, Research on Cooperatives Seminar, Kansas City, MO, April 25, 1985.

Hill, Lowell and David Timmerman. *Grain Movements to Points of Export in 1985*. Special Pub. 76, Dept. of Agri. Economics, University of Illinois.

Hineman, Kalo A., Commissioner, Commodity Futures Trading Commission, Testimony before the Committee on Agriculture, Nutrition, and Forestry of the United States Senate, September 8, 1989.

Ingalsbe, Gene, "Making Waves in Grain: Ag Processing, Inc." *Farmer Cooperatives*, USDA, September 1992.

Kimle, Kevin, Marvin, and Hayenga. "The Structure of U.S. Grain Industries," Department of Economics, Iowa State University, December 1991.

Peck, Ann and Jeffrey Williams. *An Evaluation of the Performance of the Chicago Board of Trade Wheat, Corn and Soybean Futures Contracts During the Delivery Periods from 1964-65 Through 1988-89*. Washington, D.C.: National Grain and Feed Association, 1990.

Thurston, Stanley and David Cummins. *Regional Grain Cooperatives, 1980 and 1981*. Washington, D.C.: USDA ACS Res. Rep. 27, April 1983.

"ADM and Louis Dreyfus Into Joint Venture Involving LDC Elevators," *Milling and Baking News*, October 12, 1993.

"Ag Contracts Undergo Changes in Delivery," *The Commodity Futures Professional*, 1st Quarter 1992.

"Changing Face of Breadstuffs." *Milling and Baking News*, 1983.

"Con Agra Grain, Ferruzzi Group in pact to export from Gulf," *Milling and Baking News*, November 26, 1991.

"Continental Bunge in Venture to Export from Louisiana Elevators," *Milling and Baking News*, March 24, 1992.

"Continental Grain, Harvest States in Export Venture," *Milling and Baking News*, September 1, 1992.

"Facing Up to Terrible Dilemma in Grain Trade." *Milling and Baking News*. July 2, 1991.

Farmer Cooperative Statistics, 1993, USDA, Rural Development Administration, CS Service Report 43.

"Farmland Expects Added Value from Return to Grain Trading," *Milling and Baking News,* January 18, 1994.

Grain Guide 1988. North American Grain Yearbook. Sosland Publishing Company, 1988.

"Grain Terminal Must Adapt to New Role," *Milling and Baking News,* June 5, 1984.

"GROWMARK and ADM Announce Plans for Joint Venture." GROWMARK news release, September 5, 1985.

Harvest States Annual Report 1995. Harvest States Cooperatives, St. Paul, Minnesota.

"Market Structure and Pricing Efficiency of U.S. Grain Export System," U.S. General Accounting Office, GAO/CED-82-62, June 15, 1992.

"Partnering as Newest Mode for Grain Foods," *Milling and Baking News,* June 30, 1992.

"Zen-Noh's U.S. Elevator," *Milling and Baking News,* July 5, 1983.

6

The Case for Public Grain Grades

Lowell D. Hill

The Purposes of Grades and Standards for Grain

Traditional marketing texts list grades and standards as an example of a facilitating function, i.e., one which increases the efficiency and effectiveness of marketing transactions. Grades and their associated descriptive terminology provide information to enable buyers and sellers to identify value without visual inspection. When U.S. grain first started moving in the international markets, quality was described with terms such as "sweet," "merchantable," and "sail grade." (Hill, 1990, p. 14]. As grain markets became more complex, more widely separated over time and space, and more sophisticated, the need to establish value by simple descriptive terms became increasingly important. The vague terms of the 1800s were replaced by more quantitative measures. The increased ability to describe grain quality has been matched with improvements in communication. Advances in communication and measurement technology have accompanied the development of objective, quantitative grades, until the present situation in which multi-million dollar sales can be consummated with fax or e-mail.

When Congress passed the 1916 Grain Standards Act (GSA), they specified the purposes of grades and standards in very general terms. The standards were developed "with the objectives that grain may be marketed in an orderly manner and that trading in grain may be facilitated." (USDA, 1916) H. J. Besley of the USDA's Grain Division, Bureau of Agricultural Economics, emphasized the facilitating function of the grades in 1922: "The primary purpose of grain standards, uniform throughout the country, is to facilitate trading between buyers and sellers who are distances apart . . . grades provide . . . a common language . . . grades for grain as a whole must be broad enough to cover the range of quality and condition of the commodity commonly found in commerce." (Besley, 1925). This broadly stated purpose provided the rationale for establishing grades for all major grains and for more than 300 changes in grades and nomenclature between 1916 and 1986, at which time the Grain Quality Improvement Act introduced economic justification into the purposes of grades and standards. The 1986 legislation was influenced by the Grain Quality Workshops[1] in their publication "Commitment to Quality." The 1986 amendment to the GSA introduced four specific purposes for grades and standards. The purposes are: (1) define uniform and accepted descriptive terms to facilitate trade; (2) provide information to aid in determining grain storability; (3) offer end users the best possible information from which to determine end-product yield and quality; (4) create the tools for the market to establish quality-improvement incentives ("Commitment to Quality," 1986).

Congress continued to express concern about the effect of quality on U.S. competitiveness in the export market, and in the 1990 Farm Bill they again amended the GSA to add two more purposes for grades and standards. Grades "(1) shall reflect the economic value-based characteristics in the intermediate and end-uses of grain, and (2) shall accommodate scientific advances in testing and new knowledge concerning factors related to, or highly correlated with, the end-use performance of grain" (U. S. Congress, 1990). The six purposes now included as amendments to the U.S. Grain Standards Act are an extension of the underlying objective of grain grades: to communicate information about value to the numerous players in the domestic and international markets for grain.

Official grades and standards have provided international traders with a language enabling grain to be purchased from U.S. origins with confidence that the quality loaded on the vessels will meet the specifications in the grades. Although many buyers and processors are interested in more information than the grades provide, they know that the required federal inspection guarantees minimum quality on those factors included in the grades.

Domestic processors use grades to control uniformity in the physical attributes of the grain that they receive. However, most processors are interested in additional attributes that influence the quantity and quality of products derived from the grains and oilseeds which they purchase. Consequently the official grades are often supplemented with other characteristics. Processors are increasingly using contracts that specify quality attributes including variety, cultural practices, harvesting method, and storage and drying technology. Foreign buyers are including additional quality specifications (such as oil and starch contents, and percent of stress-cracked kernels) in their contracts. Over 60 percent of soybean export inspections include oil and protein contents in addition to numerical grade (Plaus, 1994).

Country elevators use grades to assure inbound and outbound quality control, permitting them to establish a merchandizing margin that is seldom affected by differences in quality. The primary concern of the country elevator is to maintain the quality between receipt and shipment to avoid discounts. One of the functions of the country elevator is to provide a standard, uniform product to the next stage in the market channel, despite the variability in quality received from farmers. The payment for that service is often referred to as "blending income."

Processors' quality preferences are communicated to producers through two primary market signals. Buyers set discounts for attribute quality less than the base grade. These discounts are usually based on factors and limits in the official grades, although some buyers are including additional factors in their pricing scheme. Market signals are also transmitted by contracts between processors and producers, in which quality standard and price differentials are explicit, and the identity of each shipment is preserved from the producer to final user. These contracts are limited to high-value products such as food quality corn and soybeans, because it is an expensive alternative to generic commodities.

Structure of the U.S. System of Grades and Standards

The 1916 Grain Standards Act was a compromise between those who wanted the grading system to be under complete federal control from the farm to final user and those who wanted grading and inspection under control of private firms. This compromise created a system of nationally uniform grades and standards under the control of a federal agency, but left the responsibility for inspection and grading with private agencies. The private agencies were licensed to conduct inspections at domestic and export points.

Under the 1916 GSA, all grain moving in interstate commerce was required to have official grades. However, intrastate shipments and purchases from farmers did not require official grades, and inspecting and grading could be performed by employees of the firm. In addition, the 1916 GSA provided for a two-tier system of appeals. If buyer or seller questioned the results of a licensed inspector the results could be appealed to a regional office of U.S. Department of Agriculture. If that result was still questioned, an ultimate appeal could be made to the Board of Review of the USDA.

In 1968 the Grain Standards Act was amended to remove the requirement that grain in interstate commerce had to be officially inspected. The inefficiencies involved in moving grain into inspection points prior to crossing state lines had added unnecessary costs to the grain markets for many years. The state line was an artificial demarkation for official inspections, especially when the grain was moving between two plants of the same firm. This amendment allowed individual firms to choose between the more costly official inspection and grade, and the less expensive alternative of "house grades."

In 1975 a major scandal at the ports resulted in revisions in the GSA and a reorganization of the U.S. system of grain grades and standards. Following numerous indictments and convictions on fraud, altered grades, and improper weighing, Congress created the Federal Grain Inspection Service (FGIS). This agency was given responsibility for research and development related to grades and standards, implementation and enforcement of grade-related regulations, and supervision of the licensed inspection agencies. Private agencies continued to grade domestic grain shipments, but the 1976 amendment required that all grain for export be graded by a federal employee. The balance between private and government responsibility created by the compromise of the 1916 GSA was shifted to give the federal government a larger role, in order to restore credibility to the system.

In 1994 another reorganization took place. In response to continuing pressure to reduce administrative costs and streamline USDA, the Packers and Stockyards Administration was combined with the Federal Grain

Inspection Service to create the Grain Inspection, Packers and Stock-yards Administration (GIPSA) to administer the programs and functions of both agencies. The Secretary's memorandum of 20 October 1994, stated that "the results will permit us to deliver programs and services to the publics we serve in the most efficient and cost-effective manner possible" (USDA, 1994).

Structure of U.S. Grades

The purposes of nationally uniform grades require that the factors have economic importance and the grades be structured to provide information quickly and in simple nomenclature. Measures of quality within the official grading system fall into three categories: (1) grade-determining factors, (2) non-grade standards, and (3) informational factors (referred to as official criteria in FGIS terminology). **Grade factors** are those attributes that are used in the determination of numerical grades. Maximums or minimums on these factors determine whether the sample falls into No. 1, No. 2, or No. 3, etc. Every official grade certificate must include information on all grade factors. **Non-grade standards** are those attributes that are required to be measured on every official inspection but do not determine grade. For example, moisture content is no longer a grade-determining factor (removed from corn, soybeans and grain sorghum in 1985 and from wheat in 1934) (Hill, 1990, p. 128). While these factors must be measured whenever an official grade is taken, their value does not enter into classification of the sample into one of the numerical grades. **Informational factors** are attributes for which the FGIS has established standards and measurement technology but which are measured only upon request of the buyer or seller. For example, oil and protein contents of soybeans and protein content of wheat are available upon request but are not measured on every sample. Including that information on the contract is optional.

USDA has never established clear criteria to determine which factors should be placed in each of the three categories. For example, moisture was changed from a grade factor to a non-grade standard in wheat in 1934, but was retained as a grade determining factor for corn, soybeans, and sorghum until 1985. In evaluating past and future changes in grades it would be useful to have criteria on which to base the designation of an attribute as grade determining, non-grade standard, or informational factor. The following criteria provide a point of departure for continued discussion by industry and government regulators as they review future changes in grades.

Grade-determining factors should be those which are considered to be defects in the grain, where less is always preferred to more, and where zero is the most desirable value. Damage and foreign material would therefore be grade-determining factors; moisture would not because zero is not the most desirable value.

Non-grade standards should be those characteristics which are important to the majority of users in the market place, but where the exact level may differ with time, location, and use. For example, moisture content is important to all buyers and sellers of grain but cattle feeders may prefer a higher moisture content than exporters. Protein is another example of a non-grade standard. The optimum protein content of wheat differs with end use.

Informational factors are those that are important to some users and for which standardized methodology and equipment would increase the reliability of the information available. However, factors used by only a small number of market participants do not justify a measurement on every sample. For example, hilum color on soybeans is important to tofu manufacturers but it would be inefficient to require all soybean processors and handlers to identify hilum color as a basis for pricing. Breakage susceptibility and hardness are important attributes for corn dry millers, but since this industry uses less than 10 percent of the total crop, it would be economically irrational to require that these attributes be measured on corn intended for livestock feed.

Factors to be measured and factor limits for each grade have been changed many times in the history of the U.S. grades. Many of these changes have been reversed more than once, sometimes going through two or three changes only to be returned to the original level. This suggests a need for objective criteria to use when selecting grade factors and in setting limits which place a sample in one of the numerical grades. The factors to be included in any of the three categories should be based on their economic importance for the primary users. However, the value in use was not included as a criterion when the original grade factors were selected. Despite numerous changes in the details of grades, their general structures has been retained. There is strong pressure for "status quo," and the current set of grade factors for each grain are similar to those of 50 to 80 years ago when grades for the different grains were promulgated. However, the introduction of end-use property in the 1986

Grain Quality Improvement Act and in the 1990 Farm Bill have generated an opportunity to use an economic justification for the factors included in the grades. The transition from the directive to "facilitate marketing" to the objective "to reflect value in end use" can be illustrated by following the changes in soybean standards through their 74-year history.

Soybean Grades in Transition

When tentative grades for soybeans were issued in 1924, only a few were being processed, none were being exported, and their most important use was as a forage crop. In fact, grades for soybeans were under the Hay, Feed, and Seed Division of USDA rather than the Grain Division. The characteristics included in the grades were very similar to those of corn— moisture, test weight, heat damage, total damage, and foreign material. In fact, the test weight limits for No. 1, No. 2, and No. 3 grade soybeans were identical to the test weight limits for corn even though the legal weight per bushel for corn and soybeans was 56 and 60 lbs respectively. Several minor changes were made in soybean grades between 1924 and 1940, and questions were raised about the relationship between grade factor and the yield of oil and meal. However, the measurement of constituents was rejected on the grounds that the technology was too complicated for use at country elevators.

In 1940 grading responsibility for soybeans was transferred to the Grain Division with an amendment to the GSA. Between 1940 and 1985, 32 changes in definitions and grade limits were introduced into soybean grades. The circularity in the sequencing of some of the changes reflected the lack of definitive criteria and the continuing search for factors that would indicate the potential yield of oil and meal. For example, the grade factor of foreign material in the original grades was separated in 1941 into dockage and foreign material; this also required a change in grade limits. In 1949, the two factors were combined again into foreign material and the limits were changed again. Then in 1955 the limits on foreign material were changed again returning the grade limits to those established in 1941. The definition of damage provides another example of the lack of consistent application of criteria in setting grades. The percent of purple mottled soybeans was added to the definition of damage in 1951 only to have the limits removed from grades in 1994 following a study that demonstrated no relationship between purple mottling and quantity and quality of processed products (*Federal Register*, 1994, p. 10571; Sinclair, and Hill, 1987).

The concept of selecting grade factors that describe processing quality dates back to the beginning of soybean grades. Progress toward the goal has been slow. The history of measuring oil and protein illustrates the uneasy balance between government mandates and industry's proclivity for the status quo. Exporters and foreign buyers requested information on oil and protein contents as early as 1932, but USDA responded ". . . we are not authorized to make such tests for commercial agencies as part of the Federal Soybean Inspection Service" (Barr,1932). Research to develop methods for rapid determination of oil and protein contents was initiated by USDA's Grain Division and in 1954 they reported "the work with soybeans and flaxseed has been completed and conversion charts . . . have been prepared" (Phillips, 1938). Despite the research and survey work showing the importance of oil and protein contents and the lack of significant correlations between grade and processing yield, proposals to include oil and protein measurements in soybean grades were opposed by processors and marketing firms.

Foreign buyers used their analytical records to demonstrate that U.S. soybeans contained less oil and protein than soybeans received from Argentina and Brazil. With the loss of market share (the result of many factors other than quality) FGIS came under additional pressure to provide information on oil and protein contents.

The development of new rapid testing technology, pressure from foreign buyers, and the end use requirement introduced into the GSA in 1986, persuaded FGIS to offer oil and protein tests on request starting in 1989. Although oil and protein information was available to domestic as well as foreign buyers, very few domestic processors requested the information, and the market continued to resist pricing on the basis of levels of chemical constituents— i.e., component pricing. A proposal to introduce oil and protein contents as non-grade standards or grade-determining factors failed in the 1994 revisions in soybean grades. FGIS reported that "many in the industry are satisfied with the upon-request status of the tests." Although FGIS also reported that they received 58 comments supporting the change compared to 28 opposing mandatory testing they concluded, "Mandatory testing would place an unnecessary burden on the inspection systems" (*Federal Register*, 1994, p. 10572).

Resistance to component pricing came from several sources. Farmers argued that this would only provide buyers another opportunity to discount prices. Elevators argued that it would increase their costs of measurement and segregation. Processors argued that they paid full value for the entire crop based on the yield of oil and meal and that component pricing would not change the average price to all farmers. Component pricing might change the prices among individual farmers or among geographical locations, but it would not change the total value or the average price of the entire crop.

Competitive markets force processing margins to reflect demand and supply of processing capacity. Profits in the industry are determined by competition, and prices paid for soybeans reflect market-determined prices for oil and meal, minus processing margins. Processors pay on average, what the market has determined soybeans are worth. However, by not pricing on the basis of the individual constituents the signal being sent to farmers and plant breeders is "ignore oil and protein contents and concentrate on yield."

A few merchandisers recognized opportunities in the international market for contracts which guarantee higher oil or protein contents. Some foreign buyers experimented with contracts containing price differentials for oil and protein contents. Since individual barge loads vary in oil and protein contents, contract specifications could be met by selecting those barges meeting contract specifications, thereby generating a premium. Those barges with lower oil and protein contents could still be sold on the export market at full price for No. 2 beans.

Concurrent with the increased interest in component pricing, significant advances were being made in the technology for measuring oil and protein. The introduction of whole grain analyzers increased the speed and reliability of the analysis, and the instruments spread from processing laboratories of soybean processors to the commercial trade. FGIS approved the instrument for official inspections, and within three years of offering inspection services over 60 percent of exports had been tested for oil and protein at the request of buyers or sellers (Plaus, 1994). The domestic trade has not adopted the tests as a basis for pricing, despite the evidence that significant differences exist in value from lot to lot. However, with a state subsidy, more than 40 country elevators in Iowa have installed near infer red (NIR) whole grain analyzers in an experiment to test and segregate on the basis of composition (Coffman, 1996).

An Illinois study reported a wide range in values in the 1993–94 crop year. The range in values among geographical regions in the state was 13 cents per bushel. The range in values between the highest and lowest counties was 47 cents per bushel. The range among shipments within a county was 68 cents per bushel (Hill et al., 1995). These differences in value exist even though plant breeders have had no incentive to select for higher oil and protein. Component pricing could have a significant effect on channeling the highest valued product into its highest valued use and in encouraging the development of varieties with higher oil and protein contents. Official grades have provided a tool which the market has used to create price incentives to control moisture, test weight, splits, and damage. It has not provided an equally convenient method for differentiating on the basis of oil and protein. Many people in the industry have stated that the component information can be provided as needed by private firms—processors, inspection agencies, etc.—and that there is no justification for government "interference" in the market. This provides a focus for the debate about the role of government in creating and implementing grades and inspection procedures.

Public Versus Private Responsibility for Grades

The debate about public versus private responsibility for grades and standards dates back to the turn of the century, when industry and Congress vacillated between voluntary and mandatory national uniformity. Ten bills proposing various forms of grading legislation were submitted in Congress between 1907 and 1910 (Hill, 1990, p.60). The argument that quality was an issue to be decided in the marketplace was countered by the logic that 25 years of work by the Grain Dealers National Association had failed to gain acceptance of uniform standards by their membership, and that the confusion caused by numerous diverse grades was hurting the industry at home and abroad. In 1908, President Theodore Roosevelt joined the debate over uniform grain grading. His message to both houses at the beginning of the first session of the 60th Congress included this call for action:

> The grain-producing industry of the country, one of the most important in the United States, deserves special consideration at the hands of Congress. Our grain is sold almost exclusively by grades. To secure satisfactory results in our home markets and to facilitate our trade abroad, these grades should approximate the highest degree of uniformity and certainty. The present diverse methods of inspection and grading throughout the country under different laws and boards result in confusion and lack of uniformity, destroying that confidence which is necessary for healthful trade. Com-

plaints against the present methods have continued for years and they are growing in volume and intensity, not only in this country but abroad. I therefore suggest to the Congress the advisability of a national system of inspection and grading of grain entering into interstate and foreign commerce as a remedy for the present evils (U.S. Congress, 1908).

The U.S. Grain Standards Act of 1916 was created after 25 years of heated debate on the virtue of private independent grades versus nationally uniform grades. At the time of this debate there were 338 names or grade titles in use (Shanahan, 1907). The confusion, inconsistency, and inequities were decried by producers, marketing firms, and government agencies. Despite repeated efforts of trade associations, national uniformity was not achieved. Repeated efforts to create a voluntary program of uniform grades were doomed to failure because the voluntary option meant that each individual firm or local agency was free to alter the grades, the grade factors, and the limits. The cost of identifying true value, the inefficiencies in market transactions, and the loss of markets due to dissatisfied customers eventually persuaded even the Grain Dealers National Association to shift its support to the compromise bill requiring federal grades and standards, but allowing private inspection agencies which were to be supervised and licensed by the federal government. This compromise came back to haunt the industry during the 1975 scandal at the ports, and the 1976 amendment to the GSA required that federal employees replace all private inspection agencies in grading all grain for export.[2]

Supporters of federal standards and inspection advanced several arguments for federal involvement in grain grading during the debate on the 1916 legislation. They claimed that federal standards would give permanency and stability to grain grades, increase commercial honesty, and lead to a more uniform pricing of farm products. They also claimed that federal grades would eliminate many dishonest practices in grading and pricing that worked to the disadvantage of the farmer. For example, uniformly applied federal standards would prevent warehousemen from tough grading early in the season when farmers were making deliveries and then relaxed standards when grain marketing firms were moving grain out of their warehouses; would require grain to be graded by the same standards when going in and out of storage; and would improve the quality of U.S. export grain by preventing exporters from misrepresenting quality. The standards, if set by the authority of the United States, would give U.S. grain a better image and reputation in world markets, thus rebuilding the international market for U.S. grain. The proponents also believed that federal grading would end the practice of terminal elevators dictating grades to shipping firms without regard for the intrinsic value of the grain being shipped (U.S. Congress, 1908, p. 84-85). The strongest argument for federal grades during the debates in the late 1800s, as well as at present, was third party impartiality.

By placing grading under the control of an impartial third party (the Secretary of Agriculture), the framers of the 1916 GSA felt confident that all parties would receive fair and equitable treatment; inspectors appointed or licensed by the federal government would not be obligated to the grain trade and would owe no debt of gratitude to buyers or sellers. The use of federal employees would eliminate potential conflicts of interest when inspectors were hired by the same firms having their grain inspected. The concerns voiced by early proponents of an all-federal system proved to be justified, with the grading scandal that surfaced at export points in 1975. The 1976 amendments to the GSA inserted closer supervision and federal employees to tip the balance between private and government control of the inspection process toward public control.

The requirement for independent, third party determination of quality is being eroded with the move to require user fees for grain inspection. With the increased pressure on all government agencies to make users pay the cost of services, FGIS fees were incrementally increased during the early 1990s. As fees increased, utilization of official services declined. Overhead costs then had to be spread over a lower volume, increasing per unit cost. Congressional proposals to require even the research and standardization activities to be included in user fees were met with strong opposition from industry. The Grain Quality Workshop, representing a wide spectrum of the industry, strongly opposed funding standardization activities with user fees.

It has been estimated that present user fees would have to increase by approximately 22 percent to cover the cost of FGIS' standardization activities. A fee increase of this magnitude will add significant momentum to the trend of the domestic grain handling industry moving away from official grades to unofficial service of comparable quality but much lower cost. This trend has been evident for some time and new increased fees will only further undermine the financial resources of FGIS. Additionally, as domestic facilities are forced by competitive pressures to seek other alternatives to the official system, more of the financial burden will fall on the export industry, which is forced by law to use the official system. Therefore, increased user fees will effectively place a tax on exports at a time when the entire agricultural chain is working to expand exports of U.S. agricultural products.

> Importantly, we believe that the net result of new user fees will be to undermine the viability of the official system and reverse decades of work that has resulted in the finest system in the world for measuring grain quality (McCoy, 1992).

The increases in fees have continued to encourage users to shift to house grades or other sources of quality information. As the number of official inspections decreases so does the revenue, thereby increasing the average fixed cost. As average fixed costs increase, the user fees must also be increased to continue to cover the congressionally mandated proportion of the budget. This in turn further reduces the volume of inspections in a continuing downward spiral of higher prices and lower usage. Alternative sources of quality information may use the same technology and equipment as the official inspection agency, but credibility must be based on trust between buyer and seller. When buyer and seller are part of the same firm, the concern over accuracy is minimized. However, when there are many buyers and sellers, the requirement of prior experience between parties to the transactions increases the barriers to entry, increases transaction costs, and decreases competition in the market. Uniform grades allow transactions based on descriptions of standardized products, but they must be used in the majority of market transactions in order for the market to benefit from the economies of scale in providing information. Quality assurance based on the reputation of an individual firm allows for monopoly profits through product differentiation. The individual firm may benefit from product differentiation, but this benefit is more than offset by the costs and inefficiencies created in the market.

It has been argued that the marketplace can take care of quality differentials and that private inspections and house grades are adequate. Any firm who improperly or inaccurately reports quality will be punished by the market. This requires a careful evaluation of public versus private goods. The benefit of uniformity in the market channel is clearly recognized; the fact that No. 2 corn means the same in every year at any port has provided U.S. exporters a comparative advantage over countries that use a Fair Average Quality system with floating standards for export. Standards of weights and measures are almost universally recognized as public goods. Individual grocery stores cannot be allowed to determine the size of a pound, or dry goods stores to determine the length of a yard, or grain firms to determine the size of a ton or a bushel. Yet most of the quality measurements are as important in determining value as the scale weight. For example, the moisture content of corn is usually used to adjust the scale weight to determine the final quantity on which the seller will be paid. A 1 percent error in the moisture measurement has a greater effect on value than a 1 percent error in the scales.[3]

A review of history between 1840 and 1916 shows the difficulty of allowing each firm or each association or each state to set its own grades and standards and to do its own inspection. The argument for nationally uniform standards seems to be quite clear; the argument for federal involvement in inspection and grading is not.

The question of who should do the inspection and grading once the standards have been set and the technology put in place is more difficult. It must be answered on the criteria of credibility, independence, and objectivity. Given that many of the grade factors are subjective (e.g.,"commercially objectionable foreign odor" and color differences among kernels), the objectivity question becomes more important. The buyer may detect a different odor than the seller. Repeated moisture measurements on the same sample will produce a probability distribution approximating a normal curve. Without third party supervision the decision as to how many times to repeat the measurement and which reading to take as final may be used to the benefit of the firm doing the testing. The investigations and indictments in 1975 demonstrate the danger of relying on private firms to give objective third person measurement information.[4]

The market will discipline those firms that consistently give less than contract quality, but it may require several multi-million dollar transactions to build a reputation—good or bad. The reputation of an individual firm may well become a barrier to entry. The cost of experimenting with a new firm whose accuracy in grading is not known or whose equipment is not accurately calibrated is too great for a purchase of a two-million-bushel vessel worth as much as $12 million. Without assurance of accurate measurements, which are the same for every exporter regardless of its size and reputation, foreign buyers are not likely to experiment with a new competitor in the market. Newcomers will be at a distinct disadvantage in trying to promote their particular set of grades without the objectivity of the Federal Grain Inspection Service. Individual firms have much tivity of the Federal Grain Inspection Service. Individual firms have much to gain by product differentiation. Standardization is not their primary goal and in fact is part of the reason many marketing firms have been unwilling to support further development of grades and standards. They would prefer to have their own unique and differentiated product which can provide a barrier to entry to other firms whose reputation has not yet been developed.

Grades and standards are a public good. Inspection by an objective third person agency with no interest in the financial transactions is essential to credibility. Beneficiaries of uniform grades include farmers, marketing firms, processors, consumers, exporters, and foreign buyers. User fees place the primary financial burden on exporters by virtue of the mandatory export inspections; domestic firms can (and do) shift to house grades and independent agencies and laboratories. User fees that move the U.S. grain inspection system back toward the chaos of the 1860s will add cost to the transactions, decrease efficiency, decrease our attractiveness to foreign buyers, and further increase inequities among marketing firms of different sizes.

References

Barr, J.E., 1932. "Marketing Soybeans Basis U.S. Standards," Address delivered at the Annual Meeting of the American Soybean Association, Washington, DC, 3 September 1932, p. 3.

Besley, H.J., 1925. "Purpose of Grain Standards." *Standardization of Grading of Grain*, comp. E. G. Boerner and C. L. Phillips, (Washington, DC): U.S. Government Printing Office), p. 2.

Coffman, Bob. "Turning Corn Inside Out." *Top Producer* (Feb 1996): A-8.

Hill, Lowell D., Karen Bender, Stacy Crawford, and Dennis Zeedyk, 1995. "Soybean Quality in Illinois." Department of Agricultural Economics, Illinois Agricultural Experiment Station, University of Illinois at Urbana-Champaign. AE-4709, September.

Hill, Lowell D., 1990. *Grain Grades and Standards: Historical Issues Shaping the Future*. University of Illinois Press, Urbana, Illinois.

McCoy, Steve. "Grain Quality Workshop," 1992. Correspondence from Steve McCoy, Chairman, to The Honorable Jamie L. Whitten, Chairman of the House Appropriations Committee, 29 June 1992.

Phillips, C. Louise, 1938. "Supplement to History of Grain Inspection in the United States, 1838-1936." U.S. Government Printing Office, Washington, DC, p. 12.

Plaus, Marianne, 1994. "The Demand for Oil and Protein Testing," *Component Pricing in the Soybean Industry*. Department of Agricultural Economics, Illinois Agricultural Experiment Station, University of Illinois at Urbana-Champaign. AE-4702, January.

Shanahan, John D. "Standardization of Grain," *Modern Miller*, 13 July 1907, p. 18-19, 23.

Sinclair, J.B. and L. D. Hill, 1987. "In Search of Soybean Quality," *Illinois Research*, Agricultural Experiment Station, College of Agriculture, University of Illinois at Urbana-Champaign, Summer/Fall.

Commitment to Quality: A Consensus Report of the Grain Quality Workshops. June 1986, p. 3.

Federal Register. 7 March 1994, 59(44):10571-10572. U.S. Government Printing Office, Washington, DC.

U. S. Congress, 1990. *Food, Agriculture, Conservation and Trade Act of 1990.* Title XVIII, section 1821, Washington, DC.

U.S. Congress, House Committee on Interstate and Foreign Commerce, *Hearings on H.R. 6293, 6294 and 14770, Providing for the Inspection and Grading of Grain,* 60th Cong., 1st sess., 3 March–16 April 1908 (Washington, DC: U.S. Government Printing Office, 1908), p. 72 and 84-85.

USDA, "Rules and Regulations of the Secretary of Agriculture Under the United States Grain Standards Act of August 11, 1916," Circular No. 70 (Washington: U.S. Government Printing Office, 1916), p. 3.

USDA, 1994. "Reorganization of the Department of Agriculture", Office of the Secretary, Washington, DC, October 20.

---

[1] The Grain Quality Workshops, initiated in 1985, were sponsored by the North American Grain Export Association in response to congressional threats to legislate quality standards to help regain lost export markets. Participants included representatives from all major sectors of the grain production, marketing, and utilization system. Their deliberations were transmitted to congress and FGIS in the form of published recommendations. The Workshop is continuing with expanded membership under the sponsorship of the National Grain and Feed Association.

[2] There are a few exceptions to mandatory federal inspection at export, but the volume is sufficiently small that they can be ignored in this discussion.

[3] Because moisture is calculated on the "wet basis", a 1% decrease in moisture content from 16% to 15% results in a "pencil shrink" of 1.176%. Most elevators use "approximate" shrink factors of 1.3 to 1.5%. Thus a 1% error in scales is a 1% error in weight; a 1% error in moisture determination could mean as much as a 1.5% error in weight.

[4] For a detailed description of the investigations and indictments see Hill 1990, Chapter 5.

7

Phytosanitary Standards: The Case of Pacific Northwest Wheat Shipments to China

James R. Jones, Patricia Carlson,
Lu Qu and Maurice V. Wiese

Introduction

Technical barriers in international transactions, particularly as applied to agricultural commodities and products, are often cumbersome and politically motivated. Sanitary and phytosanitary (S&P) technical barriers in particular are increasingly recognized as in need of review and scientific validation (Hillman, 1991). Plant and animal health considerations can create uncertainty or near panic in international transactions. The discovery of Karnal bunt on wheat in the U.S. Southwest and the mad cow crisis in Britain are two of the most recent and dramatic illustrations. Often regulatory actions taken to deal with a threat of contamination can result in outright quarantine or expensive control measures that reduce or even halt international trade. In certain cases such barriers may be disguised mechanisms designed to protect domestic producers from foreign competition. In other circumstances, animal and plant health issues catalyze restrictions for legitimate purposes. The challenge always is to find a technically and economically feasible way to resolve the problem. This chapter reports on an economic evaluation of possible ways to remove a quarantine imposed by China since 1973 on wheat shipments through Pacific Northwest ports.

Wheat shipments from the Pacific Northwest are barred in China by a zero tolerance phytosanitary restriction against dwarf bunt, a wheat disease commonly referred to as TCK smut because it is caused by the fungus *Tilletia controversa* Kühn. TCK smut sporadically affects winter wheat produced in the northern United States, especially in the Pacific Northwest states of Idaho, Oregon, Washington, Montana, and Utah. It also occurs in several other wheat growing areas of the world. In the Pacific Northwest it affects only a small percentage of wheat plants. Because resistant varieties, clean seed sources, and chemical fungicides are readily available, it is generally an inconsequential constraint to yield and quality (Trione, 1982; Kronstad, 1990; Sitton et. al., 1995). However, because TCK spores from infested wheat heads are dispersed easily during harvest, transport, and storage, they contaminate most wheat shipped from Pacific Northwest ports. China's quarantine, therefore, presents a marketing problem with an economic impact that goes far beyond local wheat yield and quality issues. By association with TCK-infested wheat, barley shipments from the Pacific Northwest are also prohibited by China's quarantine authorities.

Most minimum market grades and standards can be met by blending wheat from different lots or areas. It is through blending, for example, that acceptable protein, purity, market class, and moisture levels can be achieved (Hill, 1990). In the case of TCK, since China has imposed a zero tolerance standard, blending is not a solution because trace quantities of the fungus are not eliminated. In infected wheat plants, TCK spores multiply and overtake the kernel, replacing its contents with a bunt ball of 4 to 8 million spores within a thin membranous coat. Since a wheat head usually contains 25 to 30 kernels, one wheat head can potentially produce 100 to 240 million TCK spores. Traces of these spores inevitably appear on otherwise healthy grain in the grain distribution network. Grain elevators in eastern Oregon, Washington, and in most of Idaho are contaminated by TCK spores (Mathre, 1983; Grey et al., 1986). There are so many spores produced within infected heads, and the techniques to microscopically detect spores are so effective, that one smutted head per acre is sufficient to yield grain identified by the Chinese as TCK infested.

China replaced the former Soviet Union as the world's largest wheat importer in the 1990s. China's importance as an importer is expected to continue or even expand in the next century. A study by the Organization for Economic Cooperation and Development (OECD) in 1995 projects China's wheat imports to reach as high as 40 million metric tons by the turn of this century. The Pacific Northwest is highly dependent on such export markets for its wheat. In the industry's view, lack of access to the People's Republic of China has necessitated finding wheat export markets elsewhere. Currently, major importers of Pacific Northwest wheat include Japan, Korea, Taiwan, Pakistan, Egypt, and Bangladesh. Wheat shipped to China from Portland would have lower transportation costs than wheat shipped to Bangladesh, Pakistan, and Egypt. The shipping distances to Bangladesh, Pakistan, and Egypt are 3,498, 4,876, and 5,404 nautical miles farther, respectively, than to Shanghai, China.

From a world welfare perspective, the zero tolerance of TCK by China acts as a zero import quota on wheat from Pacific Northwest ports and distorts trade. Such binding quotas restrict imports below the quantity that otherwise would occur (Houck, 1986). Providing a way to open wheat trade with China for Pacific Northwest producers would increase regional wheat demand and prices (Mick, 1991).

A more recent, but similar, phytosanitary wheat trade controversy to the dwarf bunt or TCK issue has arisen from the discovery of another smut disease, Karnal bunt, in wheat in Mexico in the early 1980s and in the U.S. Southwest in 1996. Like TCK smut, Karnal bunt is caused by a relatively persistent soil and seed borne fungus, *Tilletia indica,* that is of little biological consequence to wheat production or quality. Known for years to exist in India, Pakistan, and Mexico, over 30 nations, including the United States, limit the importation of Karnal bunt by imposing zero tolerance quarantines against countries, states, and counties where Karnal bunt occurs or is suspected to occur. China, however, is the only country that has imposed a zero tolerance of the dwarf bunt quarantine against wheat shipped through U.S. Pacific Northwest Ports.

A technology-cost assessment study of alternative ways to ship wheat to China is summarized below. An analysis of the potential benefits and costs of removing the quarantine is reported in the next section from a spatial programming simulation of trade volumes, values, and prices in the Pacific Northwest and China with and without the quarantine.

Technology Cost Assessment

If TCK smut could be eradicated in the Pacific Northwest, the trade problem with China could be easily resolved. However, the pervasive nature of the spores has already been noted. TCK spores are protected in the bunt balls and may persist in the soil for several years until environmental conditions permit them to germinate under snow at or near the soil surface (Hoffman, 1982; Trione, 1982). For this reason, predicting the sporadic occurrence of dwarf bunt is difficult. Complete elimination of the pathogen is not a likely prospect under even the most optimistic assumptions. While the disease has been successfully managed and biologically inconsequential for several years, its sporadic success at infecting occasional plants in certain areas has given it a persistent foothold in most winter wheat areas. This leaves isolation or treatment of wheat in these specific locations as the most technically likely options for meeting China's tolerance requirement.

In a cost assessment study, a systems approach can economically evaluate alternative agronomic, logistical, and transportation arrangements to ship identity preserved TCK-free wheat to China. Schruben (1968), Minden (1968), and Casavant (1971) discussed the systems approach to marketing problems and divided the marketing system into interactive or interdependent parts for analysis. Cramer et al. (1978) analyzed the marketing of TCK-free wheat from Montana to China. The objectives were to assay Montana grain for the presence of TCK spores, identify production areas in Montana that were TCK-free, determine the costs to maintain the identity of TCK-free areas, and market TCK-free wheat from Pacific Northwest ports. The Montana study assumed TCK-free production areas could be located and isolated, and that wheat from such areas would be delivered to regional TCK-free elevators and market channels. Costs were computed assuming a bulk-marketing system, and shipping TCK-free wheat from dedicated TCK-free elevators.

This chapter focuses on producing and shipping wheat from the Palouse region in northern Idaho and eastern Washington. It goes beyond the Montana cost analysis by considering intermodal transportation and chemical treatment alternatives that are technically possible procedures to remove the TCK market barrier. Information from the literature and interviews conducted with plant pathologists, plant breeders, elevator managers, transportation operators, agribusiness operators, directors of wheat commissions, and grain merchandisers were used to design hypothetical systems to deliver TCK-free wheat from Pacific Northwest ports to China.

Figure 7.1 shows the estimated costs of alternative agronomic, distribution, and inspection procedures from production in the field to markets in China. Identity preserved TCK-free wheat shipment in intermodal containers, or from dedicated TCK-free elevators, is compared to chemical treatment to eradicate live TCK spores at the point where wheat is loaded on ocean vessels, and to the traditional bulk distribution system for exporting wheat from Idaho to China. Four alternatives are analyzed. Alternative 1, the current bulk marketing and distribution system, is used as a baseline delivery system. Alternative 2 assumes intermodel contain-

Figure 7.1 • Costs to Deliver Wheat to China from North Idaho

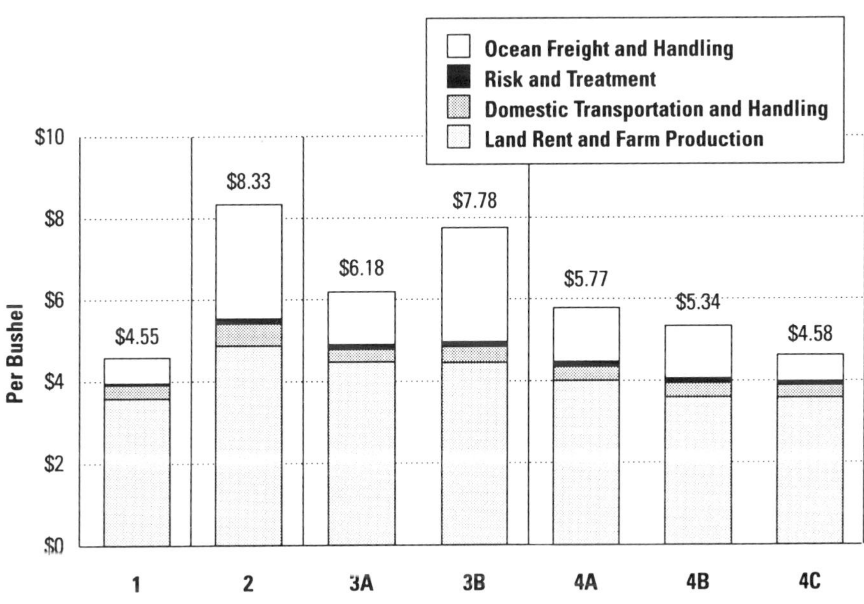

1. Current bulk production, transport, and marketing.
2. Isolation—Identity preserved, containerized, intermodal shipments.
3A. Identity preserved and bulk delivery through dedicated hinterland and export elevator.
3B. Identity preserved and containerized delivery from dedicated hinterland elevator.
4A. At-vessel sodium hypochlorite treatment.
4B. At-vessel treatment, plus new on-farm seed treatment technology.
4C. At-vessel treatment, tolerance level accepted.

ers are used to deliver wheat. Alternative 3 employs dedicated TCK-free elevators, and Alternative 4 assumes treatment to destroy viable TCK smut spores at the export elevator. Alternatives 3 and 4 were further broken into additional alternative delivery systems. In a few cases TCK-infested wheat has been detected by the Chinese in spite of the quarantine on shipments from the Pacific Northwest. When the cargoes have been accepted and treated, COFCO (formerly Ceroils), the Chinese buying agency, has applied to the U.S. shipper for compensation for the cost of the treatment. A dockage risk factor was estimated for all alternatives except the base model and treatment at the export elevator that relaxes the zero tolerance restriction to allow for possible rejection of the shipment meeting zero tolerance standards after arriving at a Chinese port. A detailed discussion of how costs were calculated is reported in Carlson, Jones, and Wiese (1993).

Alternative 1: Current Bulk Production, Transport, and Marketing.

Costs were estimated for a conventional winter wheat production and marketing system to establish a baseline for comparison. It was assumed that wheat was seeded in the fall and either stored on-farm or delivered by truck to a country elevator after harvest. The grain was shipped in bulk both inland and during the overseas carriage. The baseline delivered cost to China via Portland, Oregon for bulk wheat was estimated at $4.55 per bushel based upon cost of production estimates for the Palouse region of northern Idaho (Figure 7. 1). The costs per bushel of delivering the wheat by conventional bulk mode to Portland break down as $3.57 farm production and land rental costs, $0.30 domestic transportation and handling costs, and $0.68 ocean transportation and handling costs. While this production and delivery system is economical due to the efficiencies of scale, TCK infestation is almost inevitable due to reasons discussed above.

Alternative 2: Isolation—Identity Preserved, Containerized, Intermodal Shipments.

This alternative assumed wheat was produced in selected, TCK-free locations and transported through channels that were isolated from other contaminated grain. For this scenario, fungicide treatment of seed and soil and use of TCK-resistant wheat varieties were assumed. Harvesting, transportation, and elevator equipment were cleaned and disinfected prior to handling TCK-free wheat to avoid introducing TCK spores remaining from handling other contaminated grain. Intermodal containers were loaded with wheat in the field and ultimately transferred to general cargo container vessels for overseas transit.

The cost of shipping wheat from the production site, identity preserved with intermodal containers (Alternative 2), was estimated at $8.33 per bushel (Figure 7. 1), 83 percent more than the current production and bulk delivery system to ship wheat to China (Alternative 1). Farm production costs increased 37 percent while post-farm grain handling cost escalated 253 percent over the current bulk delivery system. While containerized domestic transportation and handling costs were expensive compared to baseline bulk handling ($ 0.50 versus $ 0.30), ocean container freight costs were the primary factor responsible for the higher cost associated with this alternative. The cost of shipping and handling containers on ocean vessels was $2.84 per bushel, compared to the baseline bulk ocean freight and handling cost of $0.68 per bushel.

Alternative 3: Isolation—Identity Preserved and Delivery through TCK-Free Elevator

In Alternative 3, wheat was produced as described in Alternative 2 but delivered through dedicated TCK-free elevators. It was assumed that a dedicated, TCK-free export facility supported by hinterland TCK-free elevators was available. Two sub-alternatives under this model were examined.

Alternative 3A: Identity Preserved and Bulk Delivery through Dedicated Hinterland and Export Elevator • With dedicated, TCK-free elevators in the hinterland and at Portland available to receive TCK-free wheat, the one-time cleaning of dedicated hinterland and export elevators were assumed to be insignificant when considered as a long-term cost. Thus, the expense for facility cleaning was not explicitly added to costs. Since TCK spores can remain in grain dust, it was assumed bulk vessel holds were cleaned and disinfested prior to loading TCK-free wheat. The total estimated delivered cost to China was $6.18 per bushel (Figure 7.1), 36 percent higher than the conventional bulk system. The increased expense to transport wheat from the production site to dedicated, hinterland elevators, cleaning of harvest and transportation equipment, and applying seed and soil treatments caused farm production costs to increase 24 percent. Post-farm activity costs were 75 percent higher than the current bulk system chiefly due to bulk vessel cleaning costs.

Alternative 3B: Identity Preserved and Containerized Delivery from Dedicated Hinterland Elevator • In this scenario, dedicated TCK-free elevators were assumed to be available in the hinterlands but not at Portland. The remaining assumptions were the same as in alternative 3A, except that transportation was by container from the hinterlands to circumvent export elevators. Estimated delivered cost to China was $7.78 per bushel (Figure 7.1), a 71 percent increase above the current bulk system. Most of the increase resulted from ocean container freight rates.

Alternative 4: At-Vessel Disinfestation

Alternative 4 couples traditional handling and bulk shipment of wheat with disinfestation treatment at the export elevator. Live TCK spores would be killed by treating wheat with sodium hypochlorite, irradiation, or oxidizing vapors, or some other technique. Research into such techniques is in initial stages and no actual TCK treatment at U.S. export facilities exists to date. The treatments outlined below assume a sodium hypochlorite treatment is administered at the export elevator spout as the ocean vessel is loaded. Cost and technical assumptions for at-vessel sodium hypochlorite treatment are extremely tentative since an actual system has not been tested to date.

Alternative 4A: At-Vessel Sodium Hypochlorite Treatment • It was assumed that wheat was sprayed with a light mist of 1.5 percent to 2.0 percent sodium hypochlorite prior to loading into the vessel. This alternative assumes the bulk grain vessel was cleaned.

When treatments to disinfest wheat occur at the export elevator, the costs were substantially reduced from identity preserved Alternative 2 and Alternative 3 because existing wheat marketing channels could be used. Treating at the export elevator eliminates the need to clean harvest and transportation equipment and storage facilities, and the need to apply soil treatment and use expensive identity preserved techniques discussed in Alternatives 2 and 3. Post-farm activity costs were higher than the base model due largely to bulk vessel cleaning costs. Overall, Alternative 4A delivered TCK-free wheat to China for an estimated $5.77 per bushel, 27 percent higher than the base (Alternative 1).

Alternative 4B: At-Vessel Treatment, New Seed Treatment Technology • Alternative 4B differed from 4A in the seed treatment used. This scenario illustrated that research on control of dwarf bunt has offered potential for cost-cutting technology. In recent years, a new seed treatment fungicide has been highly effective in controlling TCK smut development in the field. This new fungicide, CIBA Dividend (difenoconazole), in most cases can eliminate dwarf bunt development. Alternative 4B assumed the availability of difenoconazole as a replacement for existing seed treatments. Since TCK-free wheat can be contaminated with airborne and residual spores remaining in harvesting, transportation, and elevator equipment, equipment and vessel treatment with sodium hypochlorite was maintained. Because zero tolerance was assumed, the dockage risk factor and cleaning of bulk vessel holds were included in the costs to meet the standard.

The use of difenoconazole seed treatment reduced estimated farm production costs to the same level as Alternative 1. The estimated delivered cost to the People's Republic of China was $5.34 per bushel for this alternative. Overall, Alternative 4B was 17 percent higher than the current delivery system (Alternative 1).

Alternative 4C: At-Vessel Treatment, Tolerance Level Accepted • Alternative 4C differs from Alternative 4B by allowing a low TCK tolerance level, but it still assumed an at-vessel treatment to kill any TCK spores that may be present on the grain. It assumed that a tolerance level at the vessel loading point is negotiated with the Chinese and that the dockage risk factor and the bulk vessel cleaning costs assumed in Alternatives 4A and 4B were eliminated. Farm production costs were as described in Alternative 1. However, post-farm activity costs increased 3 percent due to the $0.03 per bushel ($1.10 per metric ton) sodium hypochlorite treatment at the export elevator loading site. The estimated delivered cost to the People's Republic of China was $4.58 per bushel. Assuming the zero tolerance to TCK standard required by the Chinese could be modified to allow shipments under these conditions, the estimated delivered cost to China would be 0.7 percent more than current bulk production and transportation methods.

This last alternative is viable only if the United States can convince China's authorities that the risk of TCK contamination is eliminated. In the same way that treatment with difenoconazole is assumed 100 percent effective in the field, treatment of TCK-infested wheat with sodium hypochloride could also be assumed 100 percent effective. The assurance process needs to stop with the at-vessel treatment. Further testing after treatment would eventually lead to identifying a viable spore at some time. China's authorities do not scrutinize shipments in vessels from origins other than the Pacific Northwest even though some of those vessels may carry TCK spores from other ports of loading or from some past shipment of TCK-contaminated wheat. In fact China Ocean Shipping Company (COSCO), China's state controlled steamship company, currently carries wheat from Pacific Northwest ports to other country destinations which could contaminate their own vessels.

A Trade Model Simulation

This section uses simulations from a spatial equilibrium wheat trade model to analyze the economics of removing the Chinese quarantine on wheat shipments from the Pacific Northwest. A description of the model is presented in Jones et al. (1995).

Since imposing a quarantine on wheat shipments from the Pacific Northwest in 1973, China has purchased most U.S. wheat from Gulf ports. An unrestricted base solution (Table 7.1) indicated that if wheat shipments were directed solely by minimizing transportation costs, the U.S. Pacific Northwest and China would be major trading partners.

The other two simulations reported in Table 7.1 represent scenarios where there is no wheat trade between China and the Pacific Northwest (due to China's zero tolerance quarantine) and where the treatment for TCK-infested Pacific Northwest wheat reported under Alternative 4C above is imposed.

The total trade volumes of the trading countries changed very little in the various scenarios. Under unrestricted trade (no quarantine), the Pacific Northwest was the largest supplier of China's imported wheat and U.S. Pacific Northwest ports handled most of the wheat exported to China. Using data for the late 1980s, least cost simulated annual shipments between the two parties were 6,214,000 tons, 43 percent of China's total imports, 70 percent of U.S. exports to China and 96 percent of the Pacific Northwest's (PNW's) exports. Western Canada and Australia were the major competitors in the Chinese market. Together they accounted for about 39% of China's imported wheat. U.S. Gulf ports also export a considerable amount of wheat to China. When the zero tolerance quarantine is imposed on wheat shipped from the PNW to China, Australia and western Canada assumed the Pacific Northwest's market share in China and shifted their exports from the rest of the world to China. Simultaneously the Pacific Northwest turned to the other world markets to dispose of its wheat surplus. Both export volume and price decreased in the Pacific Northwest, reducing the value of wheat exports through the region's ports by 6 percent. China's import price increased from $152 to $153 per ton as a result of the quarantine.

Table 7.1 • China's TCK Quarantine and Pacific Northwest Wheat Shipments (1000 metric tons) and Prices ($ per ton).

Imports and Prices by Regions/Countries

Exports and Prices by Regions/ Countries	No Quarantine				TCK Quarantine				TCK Treatment			
	PRC[1]	ROWM[2]	TOTAL	Price	PRC[1]	ROWM[2]	TOTAL	Price	PRC[1]	ROWM[2]	TOTAL	Price
U.S. Pacific Northwest	6214	0	6124	131	0	6001	6001	126	817	5352	6169	130
U.S.Gulf	2619	12572	15191	121	2829	12419	15248	122	2657	12544	15201	121
Western Canada	1376	4733	6109	131	6125	0	6125	132	6112	0	6112	131
Australia	4264	8698	12962	150	4866	8097	12963	150	4268	8695	12963	150
ROWX[3]	0	44968	44968	139[4]	0	45588	45588	139	0	45583	45583	139[4]
TOTAL	14473	70971	85444	—	13820	72105	85925	—	13854	72174	86028	—
Price	152	147[4]	—	—	153	147[4]	—	—	152	147	—	—

[1] PRC: Peopleís Republic of China.

[2] ROWM: Importing regions of rest of the world.

[3] ROWX: Exporting regions of rest of the world.

[4] Prices of rest of the world are average prices weighted by market shares.

* Professor Jones is a Marketing Economist in the Department of Agricultural Economics and Rural Sociology, Mrs. Carlson and Mrs. Qu were former Graduate Students and Research Assistants in the Department of Agricultural Economics and Rural Sociology, and Professor Wiese is a Plant Pathologist in the Department of Plant, Soil, and Entomological Sciences at the University of Idaho.

A sensitivity analysis of trade between the PNW and China for different treatments showed that of all the technically plausible measures discussed above, only the least costly at-vessel treatment measure (Alternative 4C) seemed to be economically feasible. At any cost above $3 per ton or approximately 8 cents per bushel there would be no trade. Assuming $1.10 per ton treatment cost, equivalent to 3 cents per bushel, there would be some wheat shipments between the U.S. Pacific Northwest and China. China would purchase 6 percent of its imported wheat from the Pacific Northwest, 13 percent of the region's total exports. If TCK-infested wheat were treated at the export spout at a cost of 3 cents per bushel and the Chinese modified their current zero tolerance testing to acceptance of this treatment, the Pacific Northwest's export value would increase by 6 percent and China would pay 8.7 million dollars less for its imported wheat due to the lower wheat price in the simulation.

The People's Republic of China's zero tolerance against TCK smut has excluded wheat exports from the Pacific Northwest ports. This exclusion results in increased transportation costs to deliver wheat from other ports and the loss of access to a potentially large wheat export market for Pacific Northwest wheat producers. Continued lack of access to China's wheat market is inevitable for the Pacific Northwest unless realistic tolerance levels for TCK smut are negotiated, or economical means of delivering TCK-free wheat are found. The results reported above suggest that sanitizing shipments at interior U.S. locations will be uneconomical because of the substantial difference between bulk handling and identity preserved shipping costs. Trade will not resume unless the Chinese are willing to pay substantial premiums. To date this has not been the case.

Eradication Versus Isolation Versus Control

While diseases such as TCK smut and Karnal bunt require vigilance, attempts to eradicate existing infestations and impose quarantines on geopolitical areas cannot be defended based on our current knowledge of smut and bunt diseases and of their biological impact. Smut and bunt diseases, while important on wheat and other cereals in the past, are now effectively managed with the use of clean seed, seed treatment chemicals, resistant varieties, and other cultural practices such as crop rotation.

It has become clear that quarantines have not prevented the movement of smut or bunt fungi into new areas where susceptible plants are grown. With the widespread distribution of cereal commodities and other agricultural products in international commerce, fungal spores associated with them have ready opportunity to travel great distances. Karnal bunt spores, in addition, are efficiently disseminated by wind and such windborne movement across the U.S.–Mexican border is unaffected by imposed quarantines.

Conclusion

While dwarf bunt, also known as TCK smut, affects only a small percentage of wheat production area in the Pacific Northwest, the intrusion of TCK spores into harvesting and transportation equipment and storage facilities creates a problem that far overshadows the biological impact of TCK smut on wheat yield or quality. The TCK quarantine imposed by China on Pacific Northwest wheat exports has been in existence for 25 years. Present evidence, including the above cost analysis, provides little assurance that TCK smut can be economically eradicated to meet a zero tolerance standard or isolated by identity preserved transportation and distribution systems. Further advances in technology could conceivably break through the economic barrier of eliminating or isolating dwarf bunt, but costs would have to fall radically. Production, transportation, handling, and treatment costs to satisfy a zero tolerance standard using present technology was found to be uneconomical in the trade simulation analysis. Focusing on procedures to eliminate or destroy viable TCK spores at the export elevator spout might be a possibility with some modification of the present zero tolerance standard. Effective and low-cost treatments applied at the time the cargo is unloaded in a Chinese port might have potential. Such a post-harvest procedure might also have potential for eliminating other wheat pathogens or pests such as Karnal bunt or flag smut should they also confront zero tolerance standards. If China maintains its current zero tolerance standard against dwarf bunt spores in wheat shipments from the Pacific Northwest, options for trade are extremely limited.

In order to counter the dramatic negative market effects of an imposed quarantine, there have been attempts to eradicate TCK smut in the Pacific Northwest. The United States furthermore is now engaged in a massive and costly undertaking (estimated expenditures exceed $37 million) to: (1) identify the distribution of Karnal bunt in the United States for the purpose of defining quarantines and identifying Karnal bunt-free areas for trading partners; and (2) eradicate the disease even though it is unlikely that the fungus can be eradicated from soil. Quarantines, if imposed, should be based on current biological evidence and should result only following thorough review and mutual resolution by scientists and policy makers from all affected wheat growing countries and wheat trading partners.

References

Carlson, P., J. R. Jones, M. V. Wiese. "A Technology-Cost Assessment of Alternatives for Shipping Dwarf Bunt (TCK) Free Wheat to China from the Pacific Northwest." Research Bulletin 154, Department of Agricultural Economics and Rural Sociology, University of Idaho Experiment Station. 1993.

Casavant, Kenneth L. "An Economic Evaluation of the Competitive Position of Puget Sound Ports versus Columbia River Ports for Pacific Northwest Wheat Exports." Washington State University, Ph.D. dissertation. 1971.

Cramer, Gail L., Michael E. Murphy, and D. E. Mathre. "An Economic Analysis of Marketing Montana TCK Smut-Free Wheat in the People's Republic of China." Montana Agricultural Experiment Station, Bulletin 699. 1978.

Grey, W. E., D. E. Mathre, J. A. Hoffmann, R. L. Powelson, J. A. Fernandez, "Importance of Seedborne *Tilletia Controversa* for Infection of Winter Wheat and Its Relationship to International Commerce." *Plant Dis.* 70(2):122-125 (1986).

Hill, Lowell D. *Grain Grades and Standards: Historical Issues Shaping the Future.* Urbana: University of Illinois Press, 1990.

Hillman, Jimmye S. *Technical Barriers to Agricultural Trade.* Boulder: Westview Press, Inc., 1991.

Hoffmann, James A. "Bunt of Wheat." *Plant Dis.*, 66(11):979-986 (1982).

Houck, James P. *Elements of Agricultural Trade Policies.* New York: Macmillan Publishing Co., 1986.

Jones, James R.; Lu Qu; Kenneth L. Casavant; and Won W. Koo. "A Spatial Equilibrium Port Cargo Projection Model." *Maritime Policy and Management: The International Journal of Shipping and Port Research,* 22 (January 1995):63-80.

Kronstad, Warren E. Ph.D., Distinguished Professor, Plant Breeding and Genetics, Oregon State University, College of Agricultural Sciences, Corvallis, Oregon 97331, personal interview, August 30, 1990.

Mathre, D. E. "Presence of TCK Smut on Wheat Grain Exported from Pacific Northwest Ports in 1983," unpublished report, 1983.

Mick, Tom. "Solving TCK Problem Would Have Single Greatest Effect on Soft White Wheat Prices." *Wheat Life*, pp. 26-27 (April 1991).

Minden, Arlo J. "A Systems Approach to Planning Long-Range Marketing Programs. *Amer. J. Agr. Econ.*, 50(5):1745-1749 (December 1968).

Organization for Economic Cooperation and Development. *The Chinese Grain and Oilseeds Sectors: Major Changes Under Way.* Paris: OECD, 1995.

Schruben, Leonard W. "Systems Approach to Marketing Efficiency Research." *Amer. J. Agr. Econ.*, 50(5):1454-1468 (December 1968).

Sitton, Jerry; Maurice Wiese; Blair Gates, Robert Forster; Roland Line; Don Mathre; Clarence Peterson; Richard Smiley; and Jackson Waldler. "Dwarf Bunt of Winter Wheat in the Northwest." Pacific Northwest Extension Publication 489. University of Idaho, Oregon State University, Washington State University (1995).

Trione, Edward J. "Dwarf Bunt of Wheat and its Importance in International Wheat Trade." *Plant Dis.*, 66(11):1083-1088 (1982).

8

Structural Changes in the North American Flour Milling Industry

William W. Wilson

Introduction

A pervasive theme in many agricultural marketing industries is that of consolidation. The dynamic structural changes in these industries have resulted in fewer and larger firms, larger plants, and increased concentration. The purpose of this chapter is to describe structural changes in the North American flour milling industry. First, structural characteristics of the flour milling industries in each country, policies, and other competitive factors are described. Second, important fundamental changes from the U.S.–Canada Free Trade Agreement (CUSTA) and the North American Free Trade Agreement (NAFTA) are described. Apparent changes in firm level strategies in the United States, Canada, and Mexico since the CUSTA are highlighted and analyzed. Structural changes are presented from a panel of data for both the U.S. and Canadian flour milling industries from 1972 to 1990.

United States Industry Characteristics and Dynamics[1]

The structural dynamics of the wheat flour milling industry likely are typical of many industries within the agricultural marketing system. Demand for wheat flour products generally had minimal (to negative) growth. However, since the early 1980s, demand for milling has increased because of a reversal of per capita consumption trends, increasing 3 to 5 percent per year in per capita consumption. Most important is the apparent renewed recognition of the importance of wheat food products in American diets and the adoption of the *Food Pyramid* as a dietary program. In addition, since 1986, export flour sales via the Export Enhancement Program (EEP) have expanded, also reversing a negative trend. Almost 8 percent of flour output has been destined for export under some form of government assistance program. Because of these two important trends, capacity use has increased to relatively high rates. However, capacity use varies substantially across regions.

The wheat flour milling industry traditionally has been concentrated in market centers contiguous to wheat production regions, such as Minneapolis and Kansas City. In making location decisions, firms must choose between being located close to the point of wheat production (i.e., an origin mill) or close to the customer (i.e., a destination mill). A strategic advantage of the former is that the mill is not dedicated to specific customers or regions. Thus, the number of potential customers for an origin mill, all located at consumption centers and capable of receiving by rail, is large.[2] These strategic advantages have to be weighed against transportation and service advantages of locating closer to customers and shipping wheat longer distances. Traditionally, an important rail pricing mechanism referred to as "transit" partially offset location disadvantages of origin milling. However, the advantage of the transit privilege has gradually diminished since railroad deregulation in the early 1980s.

Another important transportation change that has impacted this industry is the advent of unit train technology in wheat shipments, which was adopted in the early 1980s. Of particular importance is the fact that the cost of shipping wheat by rail using this technology decreased relative to the cost of shipping flour and millfeeds. Wheat flour is not compatible with unit train technology since most receivers are only large enough to purchase in single-car shipments. In addition, flour shipments require highly specialized equipment, which have limited alternative uses (e.g., sugar) compared to covered hopper cars that can be used to ship virtually any raw unprocessed (and some processed) commodity. This technical change has the general effect of favoring a transition of the milling industry away from traditional milling centers toward destination markets.

Firms in the wheat flour milling industry originally were family owned or owned by local elevator companies integrated forward into flour milling. In addition, most were traditionally single- plant firms rather than multi-plant firms. However, in the past several decades, a number of "strategic groups" have become evident in this industry. A strategic group is a cluster of firms with common assets; consequently, strategic decisions often are parallel (Oster, 1990). These strategic groups are comprised of the following (Goldberg, 1983):

- Vertically integrated food processors (Pillsbury, Nabisco, General Mills, and International Multifoods);
- Multi-unit flour millers diversified into other grain operations (ConAgra, Cargill, ADM);
- Medium-sized firms that are primarily regional flour producers (e.g., Bay State Milling);
- Small millers with one or two mills in local market niches.

These definitions are used throughout the remainder of this chapter as well as in the tables and figures. The first group is largely comprised of food processors integrated backward for procurement purposes. The second group is primarily commodity firms with operations throughout the grain marketing system, but not integrated forward. The relative importance of each of these groups has changed. Specifically, the multi-unit grain and local niche firms have grown substantially. Each of the other two groups has decreased in relative importance. Firms that were largely food processors who through time integrated backward, most likely for strategic procurement purposes, no longer dominate the industry. Though multi-unit grain firms (i.e., those with major operations in other grain marketing sectors) were relatively less important prior to 1980, most of the growth, both in terms of new plants and acquisitions, has come from this group.

Comparison of Market Structure: United States and Canada

The structure of this industry has changed. The number of plants operating in the United States has decreased from 280 to 204 between 1974 and

1990, and the average plant capacity has nearly doubled (Table 8.1).[3] Reduction in the variance in plant capacity, indicated by the coefficient of variation, suggests a general pattern of convergence in both firm and plant size. The number of firms has decreased. The average firm capacity more than doubled during this period, and the number of plants per firm has increased from 1.7 to 2.2. Also, the percent of plants that "grain firms" (i.e., the second strategic group) operate has increased from 14 percent in 1974 to over 50 percent in 1990 and 62 percent in 1992. The percent that "vertically integrated firms" (i.e., forward as the first strategic group) operate has remained at less than 10 percent (Figure 8.1) and has declined. Similar behavior has been observed in Canada. The fundamental difference is that as early as 1981, over 50 percent of the plants in

Table 8.1 • Descriptive Flour Industry Milling Statistics

| | United States | | | Canada | | |
Plants/Firms	1974	1980	1990	1974	1980	1990
Plants						
Number	280	255	204	43	35	30
Average capacity (cwt/day)	3,541	4,212	5,937	4,446	5,763	6,253
Coefficient of variability: capacity	129	122	99	109	102	97
Firms						
Number	161	140	95	28	21	19
Average firm capacity (cwt)	6,158	7,672	12,534	2,029	8,717	9,203
Coefficient of variability: capacity	150	137	118	—	—	—
Number of mills	1.7	1.8	2.2	1.5	1.7	1.6
Multiplant (%)	37	42	58	37	51	50
Plants Operated by Firm Type (%)						
Grain	14	15	33	—	—	—
Vertically integrated	9	11	8	—	—	—

Figure 8.1 • Market Share for Grain Companies and Vertical Integrated Firms: United States

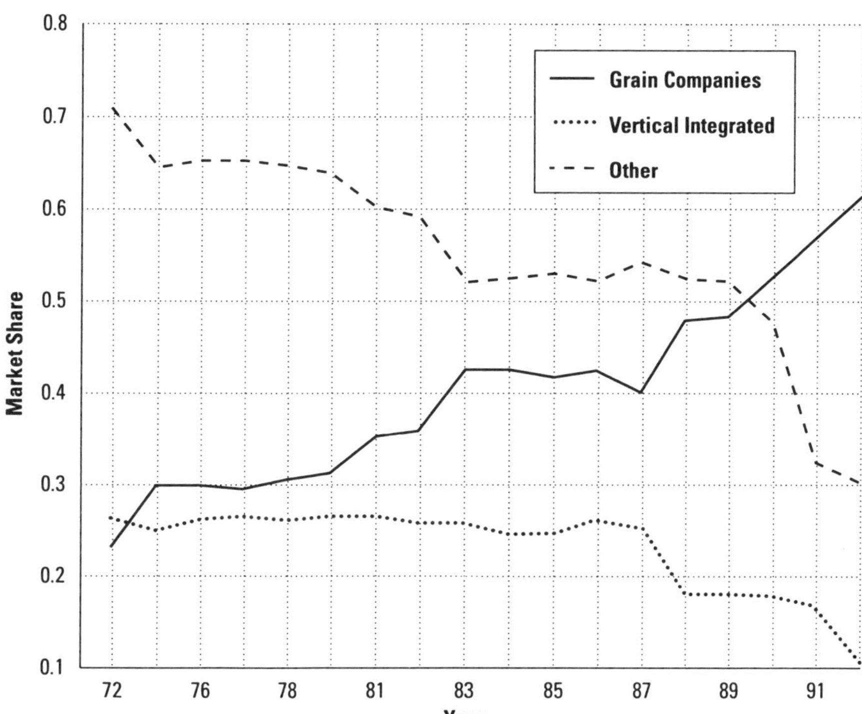

Canada were part of multi-plant operations—the U.S. industry did not surpass 50 percent until 1987.

The milling industry has become more concentrated with four-firm market (capacity) shares increasing from 34 percent in 1974 to nearly 70 percent in 1992 (Figure 8.2). The Herfindahl Index[4] reflects "balance" in the industry and is impacted by both the number and size distribution of firms. A high value of H indicates an unbalanced distribution or a greater likelihood of a dominating market leader. In the United States, H has nearly doubled since 1970.

H differs substantially across the United States. Figure 8.3 shows the Herfindahl Index for each of the Census Bureau Regions, calculated for 1972 and 1990. In all regions except Mountain, the Herfindahl Index increased through time and in most cases nearly doubled. Regions that could be subjected to fairly intense competition, reflecting a large num-

Figure 8.2 • United States Flour Milling 4-Firm Market Share and Herfindahl Index

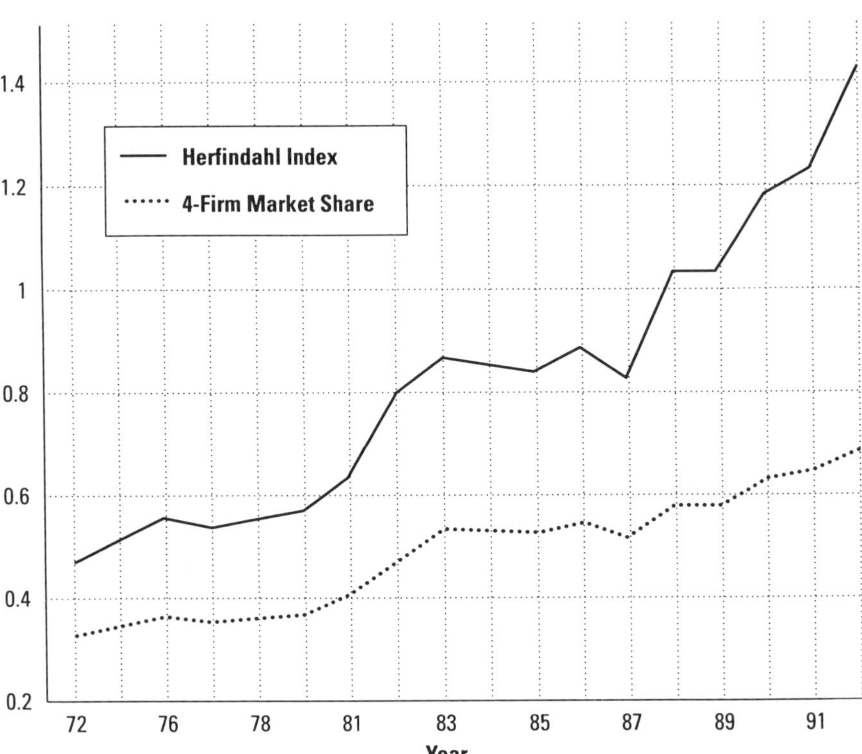

ber of similarly sized firms, are the East North Central, West North Central, and South Atlantic. In these regions, no single market leader can impact margins. The Middle Atlantic, East South Central, and Pacific are characterized more by an unbalanced competitive environment where a market leader could probably dominate. Given the identity $H=(V^2+1)/n$, the observed increase in H implies a decreased number of firms or an increased coefficient of variation in firm size (or both). However, the coefficient of variation has decreased (Table 8.1), which implies that the increase in H is attributable simply to a decreased number of firms.

Important differences in the Canadian industry are highlighted. Most important is that the Canadian flour milling industry is more concentrated. The largest three firms have controlled about 75 percent of the

Figure 8.3 • Herfindahl Index for United Sates Flour Milling Industry, by Census Bureau Regions for 1972 and 1992

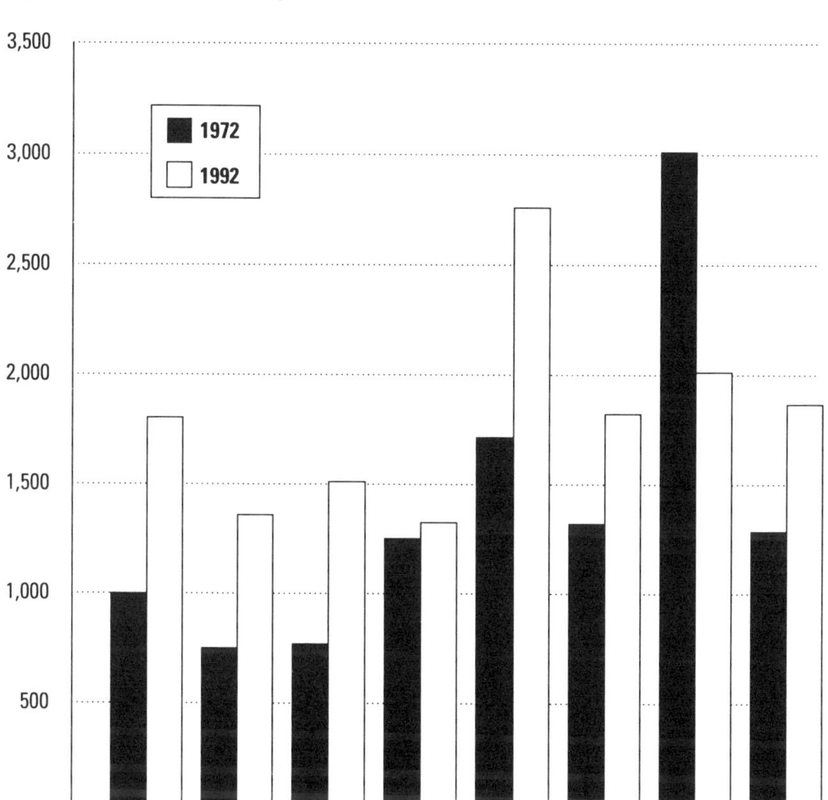

capacity since the late 1970s. The industry operates at about 70 percent of capacity. For a U.S. milling firm to be integrated into baking would be rare, but Canadian milling firms often are integrated into baking. In fact, in some regions of Canada, milling firms[0] own and operate a large proportion of the baking capacity.

The cumulative rate of entry, exit, and merger/acquisitions are shown in Figure 8.4 for the United States and Canada.[6] More detailed information about the characteristics of these is provided fin Table 8.2. For comparison, characteristics are also shown for "other" (i.e., those that did not close, exit, or merge). These results suggest several observations. First, in

Figure 8.4 • Number of Flour Milling Plants Closed, Merged, and Opened for Canada and United States

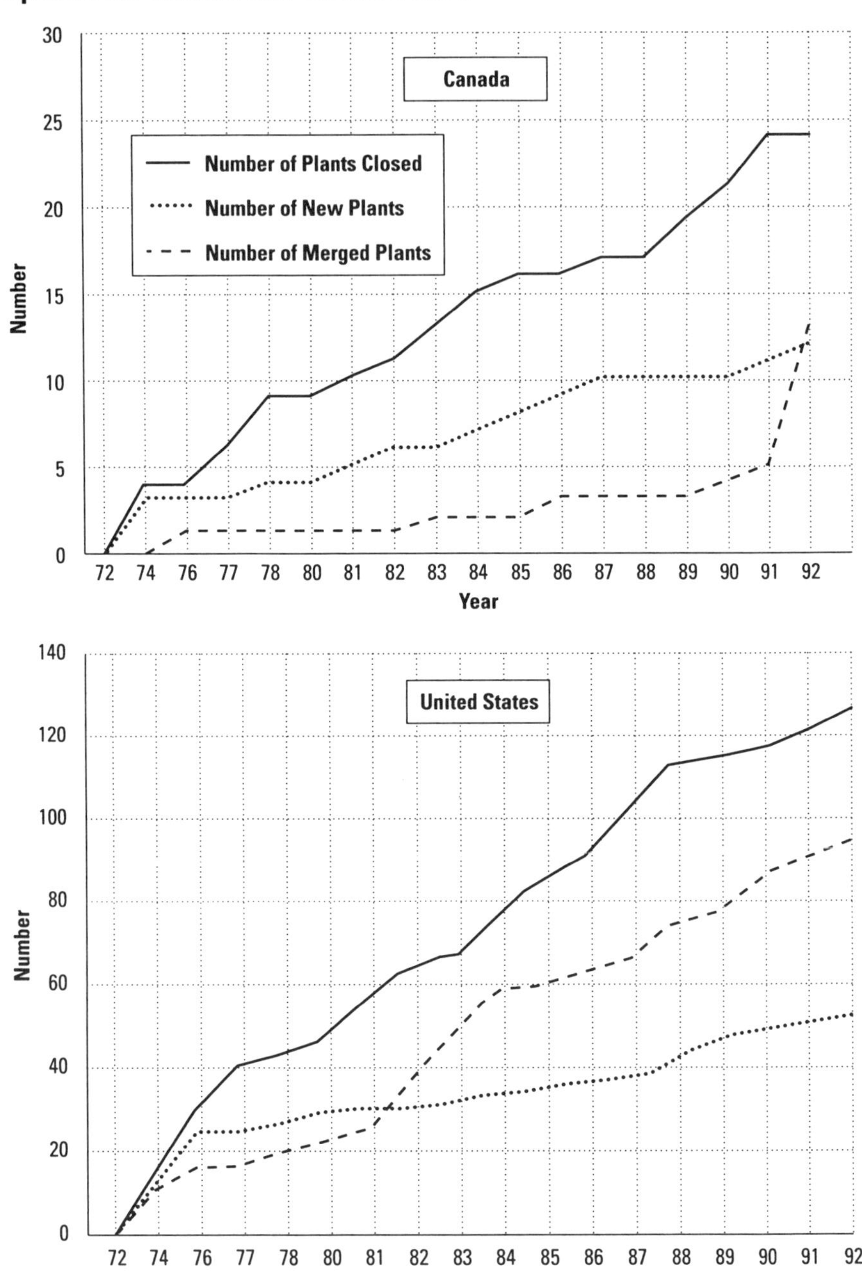

Table 8.2 • Plant Characteristics: Exited, New, and Merged (1972–1990)

Characteristics	United States					
	Exited	New	Others	Acquired Plants	Acquiring Firms	Acquired Firms
Number	112	40	—	80	—	—
Average plant capacity (cwt/day)	1,308	2,764	4,662	7,174	—	—
Percent						
Grain firms	5	13	19	61		
Vertically integrated	4	0	10	8		
Firms						
Capacity	12,383	22,074	42,011		111,657	49,195
Number of mills	2.4	3.4	5.3		13	5
Market share (Census Bureau Regions)	3	6	6		22	17
Market share (U.S.)	1	2	3		10	4
Canada						
Number	20	10	—	4	—	—
Average plant capacity (cwt/day)	2,715	4,202	5,697	6,975	—	—
Percent						
Grain firms	—	—	—	—		
Vertically integrated	—	—	—	—		
Firms						
Capacity	16,580	10,737	22,160		39,595	6,975
Number of mills	2.8	1.8	2.9		5.5	1
Market share (Census Bureau Regions)	—	—	—		25	3
Market share (U.S.)	11	7	12		20	6.5

both countries, those that exited were abnormally small for plant and firm capacity and had small market shares, regionally and nationally. Second, in the United States, existing multi-plant firms typically built new plants. However, capacity of these firms was not even close to average for the industry.

Another observation is that grain firms were the acquiring firms in 60 percent of the U.S. mergers. The average acquired plant capacity in the United States was nearly double the industry average; and in Canada, it exceeded the industry average. These observations suggest that smaller, inefficient plants exited rather than were acquired. Those plants and firms that were acquired were larger. The capacity and number of plants the acquiring firm operated were about double those of the acquired firm. The acquiring firm's market share was substantially greater than that of the acquired firm in the United States, and five times greater in Canada.

The correlation coefficients between acquiring and acquired firm characteristics in the United States are shown in Table 8.3.[7] The fact that all are positive suggests that firms in a merger have similar characteristics. This would refute the notion that large firms necessarily buy out small firms, however measured. The correlation between market shares of the two firms in the merger is among the highest. The fact that the market share correlation within the Census Bureau Regions exceeds that measured in the United States suggests that mergers were made to increase market share within a region.

Table 8.3 • Correlation of Acquired and Acquiring Firm Characteristics: United States (N=80)

	Acquiring Firms			
	Number of Plants	Capacity	Market Share (U.S.)	Market Share (CBR)
Acquired Firms				
Number	.18	.20	.16	.36*
Capacity	.28*	.30*	.27*	.40*
Market Share (U.S.)	.23*	.25*	.24*	.40*
Market Share (CBR)	.12*	.12	.13	.81*

*Indicates significant at the 10% level.

Free Trade Agreements and Flour Milling

The United States and Canada have been operating under a free trade agreement (CUSTA) since 1989, and NAFTA was recently consummated. Both of these could have important impacts on the structure of the North American flour milling industry.

U.S.–Canada Free Trade Agreement

At the time of the agreement in 1989, trade in the wheat value chain was regulated as follows: wheat and flour sales to Canada required import licenses from the Canadian Wheat Board (CWB) and were generally restrictive. Wheat-base, value-added product shipments from the United States to Canada had tariffs, ranging from zero to 10 percent. Wheat and flour shipments from Canada to the United States were subjected to tariffs, which would be reduced because of CUSTA, but benefited from subsidized rail shipments to eastern markets. Tariffs also were imposed on cereals and bakery mixes and all other products free of tariffs.

Before CUSTA, a two-price system operated in Canada, raising prices Canadian mills paid relative to third country sales. However, wheat prices were fixed, and all mills paid the same price with adjustments for transportation. Thus, mills and bakers were not able to or did not have to compete on procurement strategies. In anticipation of freer trade, the two-price system was replaced in 1989 with what was to be defined as a North American price, based on cash prices at the Minneapolis Grain Exchange.

Now mills and other end users are not treated identically throughout year. They must compete, to some extent, on pricing and procurement. The combination of these events led to major structural changes in North American milling.

North American Free Trade Agreement

In anticipation of the incompatibility of the Mexican procurement regime under NAFTA, major changes already have been adopted. Conasupo has been Mexico's import agency, in part to protect a system of domestic wheat prices that exceeded U.S. and world values. Mills were restricted to purchasing the Mexican wheat crop through an agreement called "Concurrence" at the higher domestic prices. Since this had to occur before importing wheat, storage and financing costs were inflated, and quality was highly uncertain. Bread prices were regulated, and flour was an administered function of the wheat and bread markets.

Conasupo made imports for the industry and sold to the industry at a predetermined price structure. Flour imports were unrestricted but were subject to a 15 percent tariff. Other products had either 10 percent or 15 percent tariffs. From a competitive perspective: (1) firms were treated equally in procurement, a crucial area of competition; and (2) product imports were increasingly favored relative to local processing. In fact, during the late 1980s flour and product shipments from the United States to Mexico grew rapidly.[9]

The principal feature of the proposed NAFTA regarding the milling sector is that a tariff (15 percent) would replace wheat import licenses (and, therefore, Conasupo).[8] That tariff and those on all wheat value-added products would be reduced over a 10-year period (generally). Domestic wheat production is expected to fall because of reduced prices.

Competitive Issues and Recent Structural Changes

The combined impact of these changes along with changes otherwise occurring in these industries set in motion a series of strategic changes that are redefining the North American flour industry. The apparent strategies of principal firms in each country are discussed below.

Canada

Concurrent with negotiation of the CUSTA, a merger was proposed between Ogilvie and Maple Leaf Mills, the largest milling firms in Canada, to increase scale and reduce costs and to be defensive against potential entry of U.S. firms. However, the government of Canada rejected the merger largely on antitrust grounds. The major issue was the appropriate geographic scope of competition for antitrust purposes. Almost immediately, a series of structural changes were announced, including ADM acquisition of Weston Soo Line Mills in September 1990, and Ogilvie in May 1992; then Maple Leaf Mills and ConAgra announced a venture to jointly operate their eastern mills in June 1992. The result of these cumulative changes was an increase in H in a newly defined region encompassing Eastern Canada and the contiguous U.S. regions, from less than 1,000 in 1989 to over 1,200 in 1992.

United States

Pillsbury was the largest milling firm in the early 1970s; however, following its acquisition by Grand Met, it began divesting from this industry (and apparently all other commodity-type industries). Cargill bought four Pillsbury mills for $100 million in June 1991, and ADM entered a joint venture and assumed management of the other four Pillsbury mills ($68.5 million) in March 1992. Ultimately, these were acquired by ADM.

These structural changes reveal some of the firms' strategies. First, this industry has become an industry that commodity companies dominate, a trend that began in the late 1970s. These firms have limited potential to integrate forward, which would result in their competing against their customers. Most firms in this industry are now integrated backward, thereby raising entry costs for potential new entrants. This is especially true if, as a result of integration, the firms can improve their logistical efficiency and quality control.

Second, because of these acquisitions and new ventures, firms are competing to dominate particular regions, causing the higher values of H in a number of regions. Cargill dominates California, ADM the Midwest and Pacific Northwest, and ConAgra the East.

All U.S. firms are trying to penetrate the Mexican market because of the excess capacity that exists in the south central part of the United States. Sustainability of this strategy, however, ultimately depends on whether railroads promote long hauls of flour or wheat shipments. Long haul wheat shipment has been the longer term trend elsewhere in North America.

Mexico

The largest bakery and milling company in Mexico is Bimbo, which controls a large percentage of the bakery market. Bimbo has expanded and entered into joint ventures with Keebler and Sara Lee, in both cases to distribute (Keebler and Sara Lee) products. They have built a new state-of-the-art mill to serve the Mexico City market and acquired one in Veracruz for imported wheat. Thus, both backward and forward vertical integration are important elements of longer-term strategies in Mexico. This gives a disadvantage to U.S. mills trying to penetrate into these markets.

An important component of Mexican firms' strategies is technology. The Mexican milling industry is relatively antiquated and underutilized with low levels of productivity (see Table 8.4). They have not adopted computerized control systems, quality control, or other technical improvements which have been adopted elsewhere in North America.

Mexican mill strategies likely involve improving milling and logistical efficiency for wheat imports and product distribution. Firms will undoubtedly have to improve productivity or else be disadvantaged relative to U.S. mills with excess capacity seeking to expand flour exports.[9] Besides the potential logistical advantage of wheat imports, being able to procure wheat from multiple sources is a strategic advantage. However, wheat production practices in Mexico vary in quality; mills are having to implement elaborate quality control processes to compete with imported flour.

Conclusions

A pervasive theme throughout the agricultural marketing industries is that of consolidation. A combination of firms exiting and mergers and acquisitions explain this. Understanding this dynamic evolution of industry structure is important for understanding forces that impact competition in these industries. This chapter described the dynamic evolution of the wheat flour milling industry.

The flour milling industries evolved independently in the United States, Canada, and Mexico. Policies, regulations, and the competitive structure of the vertical market system differed drastically across these countries. Exports between countries have been virtually nil, and plants generally ship within their national boundaries without being impacted directly by developments in the contiguous country. However, features of CUSTA and NAFTA are having important impacts on the structure of the industry.

Table 8.4 • Productivity Comparisons in Milling Industries Between Mexico and the United States

	Mexico	United States
Capacity Utilization	58%	85%
Average Size	40,000 mt/year	90,000 mt/year
Labor Productivity	375 mt/year	900 mt/year

This industry has four important strategic and structural characteristic. First, although the number of firms and plants has decreased, plant and firm capacity has increased. The number of multi-plant firms and firms in the "grain" strategic group also have increased. Second, the geographic boundaries among these firms are becoming blurred. However, each is gaining dominance in particular regions. Third, procurement is becoming a major element of strategy for both Canadian and Mexican mills, an area previously protected because of their agricultural policies. Finally, the extent of integration varies in the three countries: U.S. firms are largely integrated backward, and Canadian and Mexican firms integrated forward.

References

Dahl, R. "New Growth in Flour Milling," *Minnesota Agricultural Economist*, 672, 1993.

Goldberg, R. "Economics of the Flour Milling Industry," *Research in Domestic and International Agribusiness Management*, JAI Press Inc., 1983.

Harwood, J.; M.N. Leath, and W.G. Heid, Jr. *The U.S. Milling and Baking Industries*, USDA: Economic Research Service, Washington, 1989.

Krause, J.; W. Wilson, and F. Dooley. "Market Segments in Value-Added Wheat Product Trade," paper presented at the International Agricultural Management Association and meetings, San Francisco, 1993; and Ag. Econ. Rpt. #315, Department of Agricultural Economics, North Dakota State University, Fargo, 1994.

Oster, S. M. *Modern Competitive Analysis*, Oxford University Press, New York, 1990, p. 61.

Sosland, M. "Dynamic Growth in U.S. Durum Milling," *World Grain*, 11(3), pp. 6-10, 1993.

Wilson, W. and W. Wilson. "Mergers and Acquisitions in the North American Flour Milling Industry," Agricultural Economics Paper, Department of Agricultural Economics, North Dakota State University, Fargo, 1994.

Government of Canada. *Industry Profile: Flour Milling*, Industry, Science and Technology Canada, 1988.

Sosland Companies Inc.. *Milling Directory/Buyers' Guide*, Merriam, Kansas: Sosland Publishing Company, (Selected Years).

---

[1] See Harwood et al. (1989) and Dahl (1993) for recent descriptions of the U. S. milling industry, and Government of Canada[a] (1988) for description of the Canadian industry.

[2] See Sosland (1993) for a specific discussion.

[3] All firm and plant data were taken from *Milling and Baking News* annual directory.[7] Survey data were available for every year since 1972, except for 1973, 1975, and 1979.

[4] As used in this study the Herfindahl Index is defined as $H = \sum s_i^2 \cdot 10,000$, where s_i is the market (capacity) share of firm i.

[5] This is well known; however, the extent that firms are integrated forward into baking (or baking backward into milling) is very difficult to document.

[6] A companion paper (Wilson and Wilson, 1994) developed a multi-nominal model of entry, exit, and acquisition decisions in the U.S. flour milling industry. Results of that study indicated that a number of factors are important in this evolution. One of these is the high probability of exit, or being acquired, for smaller plants. The probability of acquiring another plant increases with the acquiring firm's market share. In addition, the probability of being acquired increases with the growth rate of the acquired firm's market share. This is contrary to other research, and suggests that to be acquired, the firm must have demonstrated some success. In general, multi-plant economies did not seem to be an important motive for merging not did growth in industry output or market power. Capacity use had an important positive influence on firms' entering regions with new plants. Real milling margins have declined through time, which has reduced the rate of new entrants.

[7] Due to the low number of mergers in Canada, these were not calculated.

[8] Private sector purchases already have replaced functions of Conasupo. However, a newly created agency, Aserca, which was intended to intervene during the transition period until NAFTA was ratified, has absorbed some of these functions.

[9] For reference, the idle capacity in the United States is sufficient to serve 80 percent of the Mexican market.

9

Growth and Structure of the Wet Corn Milling Industry

Paul L. Farris

Growth Characteristics

In this chapter we highlight the economic features of the wet corn milling industry, also known as the corn refining or the starch industry. Emphasis is focused primarily on two major dimensions: (1) growth and market characteristics of the industry, and (2) organizational developments and issues.

Growth of the wet corn milling industry has increased substantially in the United States, and production of starch products has expanded in several other countries. Although people continue to consume some starch directly from plants, either raw or cooked, demand for commercially produced starch to be added to food and especially beverages has increased greatly. Starch demand has also expanded for use in a wide range of industrial products, such as paper, textiles, building materials, and alcohol for fuel.

U.S. Census of Manufactures data show that the quantity of corn (including minor quantities of sorghum grain) that was processed by the wet corn milling industry increased five-fold between 1972 and 1992, from 262 million bushels to 1,303 million bushels during the 20–year period. This rapid increase took an increasing share of expanding corn production in the United States. The manufacture of wet milled products, which accounted for around 5 percent of U.S. corn production in 1960s and 1970s, fluctuated around 15 to 20 percent in the early 1990s (Table 9.1). This percentage fluctuation was associated primarily with year-to-year variations in corn production because wet milling demands for corn increased quite steadily.

Large increases in demands for two major products, high fructose corn syrup and fuel alcohol, propelled high industry growth beginning in the 1970s. The market for high fructose corn syrup was stimulated by the growing acceptance of corn sweeteners in food and especially beverage products. Prices of corn sweeteners were competitive with U.S. sugar prices, which were substantially higher than world sugar prices due to government policies (Table 9.2). Production of alcohol for engine fuel also increased greatly in the 1970s, motivated by rising crude oil prices and by government subsidies, which in the 1990s continued to exist and support the demand for fuel alcohol.

Relatively steady increases occurred in production of standard starch and syrups, along with accelerating expansion of high fructose corn syrup (HFCS) since the 1960s, and fuel alcohol in the 1980s and 1990s (Table 9.1). These two products together accounted for less than one-third of all food, seed, and industrial uses of corn in 1980-81 and for nearly 60 percent of these uses in the mid 1990s. HFCS production more than tripled, using 165 million bushels of corn in 1980–81 and 462 million bushels in 1994–95. The rise in fuel alcohol production was even more dramatic, taking 35 million bushels of corn in 1980–81 and more than 400 million bushels per year in the mid 1990s. Fuel alcohol utilized over 5 percent of total U.S. corn production during the early 1990s.

Numerous edible and industrial products are manufactured by the wet corn milling industry through further refining. The dollar value of industry product shipments more than doubled between 1982 and 1992 (Table 9.3).

High Fructose Corn Syrup Consumption

Before the accelerating expansion of HFCS, corn sweeteners, until the late 1960s, accounted for less than 15 percent of the U.S. total caloric

Table 9.1 • Corn: Food, Seed and Industrial Use, 1980–81 to 1996–97[1]

Year	HFCS	Glucose and dextrose	Starch	Alcohol		Cereals & Other Products	Seed	Total	U.S. corn production	Total FSI as % US production
				Fuel	Beverage	*(Million bushels)*				
1980–81	165	156	151	35	78	54	20	659	6,639	9.9
1981–82	183	160	146	86	86	53	19	733	8,119	9.0
1982–83	214	165	150	140	110	60	15	854	8,235	10.4
1983–84	265	167	161	160	88	70	19	930	4,174	22.3
1984–85	310	167	172	232	84	81	21	1,067	7,672	13.9
1985–86	327	169	190	271	83	93	19	1,152	8,875	13.0
1986–87	338	171	214	290	95	109	17	1,233	8,226	15.0
1987–88	358	173	226	279	85	113	17	1,252	7,131	17.6
1988–89	361	182	215	287	117	117	18	1,298	4,929	26.3
1989–90	368	193	219	321	129	120	19	1,370	7,532	18.2
1990–91	379	200	219	349	135	124	19	1,426	7,934	18.0
1991–92	392	210	225	398	161	128	20	1,534	7,475	20.5
1992–93	415	214	218	426	136	129	19	1,556	9,477	16.4
1993–94	444	223	223	458	110	131	20	1,609	6,336	25.4
1994–95	465	231	226	533	100	132	18	1,704	10,103	16.9
1995–96	482	239	219	396	110	133	20	1,598	7,374	21.7
1996–97	515	240	225	425	110	134	21	1,670	9,293	18.0

Source: U.S. Department of Agriculture, Economic Research Service, Feed Situation and Outlook Yearbook, FDS-1995 (November 1995.)
[1] Marketing year beginning September 1.

Table 9.2 • HFCS and Sugar Prices and Producer Price Index for Total Finished Goods, 1981–94.

Year	HFCS-42 Whole-sale list prices, Midwest markets, dry weight	HFCS-55 Whole-sale list prices, Midwest markets, dry weight	U.S. raw sugar prices, duty fee paid, New York	World raw sugar prices[1]	World refined sugar prices[2]	U.S. producer price index for total finished goods, 1982 = 100
	(cents per pound)					*Index*
1981	21.47	23.59	19.73	16.93	20.51	96.1
1982	14.30	18.81	19.92	8.42	11.36	100.0
1983	18.64	21.06	22.04	8.49	11.40	101.6
1984	19.94	22.69	21.74	5.18	7.71	103.7
1985	17.75	19.95	20.34	4.04	6.79	104.7
1986	18.07	19.96	20.95	6.05	8.47	103.2
1987	16.50	17.46	21.83	6.71	8.75	105.4
1988	16.47	18.68	22.12	10.17	12.01	108.0
1989	19.24	21.41	22.81	12.80	17.16	113.6
1990	19.69	21.88	23.26	12.55	17.32	119.2
1991	20.93	23.25	21.57	9.04	13.41	121.7
1992	20.70	23.00	21.31	9.09	12.39	123.2
1993	18.83	20.93	21.62	10.03	12.79	124.7
1994	20.17	22.47	22.04	12.13	15.66	125.5
1995	17.04	19.00	22.96	13.44	17.99	127.9

Source: United States Department of Agriculture, Economic Research Service, *Sugar and Sweentener Situation and Outlook Yearbook*, SSRV18N2 (June 1993) and SSSV21N4 (December 1996). U.S. Department of Labor, *Monthly Labor Review* (various issues).

[1] Contract No. 11. f.o.b. stowed Caribbean port (including Brazil) spot price.

[2] Contract No. 5. London Daily Price for refined sugar, f.o.b., Europe, spot price.

Table 9.3 • Corn Product Shipments by the Corn Wet Milling Industry. (Census Industry 2046)

Product and code	1982 million dollars	1982 percent	1987 million dollars	1987 percent	1992 million dollars	1992 percent
corn sweeteners, 20461	1610.4	51.8	2182.5	49.1	2911.0	45.4
manufactured starch, 20462	655.1	21.1	774.3	17.4	1305.5	20.3
corn oil, 20463	234.9	7.6	613.1	13.8	801.6	12.5
by-products, 20464	577.7	18.6	845.8	19.0	1363.5	21.3
not specified by kind (m.s.k.)	27.6	.9	30.5	.7	34.0	.5
Total	3105.7	100.0	4446.2	100.0	6415.5	100.0

Source: U.S. Department of Commerce, Bureau of the Census, Census of Manufacturers, (indicated years).

sweetener market. By the mid-1980s the corn sweetener share of total caloric sweeteners had risen to more than 50 percent, and the upward trend, though slowing, was continuing (Table 9.4). HFCS had rapidly replaced most other sweeteners in the non-alcoholic beverage market (Table 9.5). Beverage use accounted for nearly three-fourths of HFCS production in the mid-1990s.

Production of HFCS in other industrialized countries is far lower than in the United States, but the trend has been increasing (Table 9.6). Future significant growth is expected in other countries, but if relatively lower sugar prices prevail abroad, the pace of HFCS growth in other countries will likely be slower than past HFCS growth in the United States.

Fuel Alcohol

Search for alternative energy sources beginning with the crude oil shortage and crisis in the 1970s led to renewed emphasis on alcohol as an

Table 9.4 • Corn Sweetener Consumption Per Capita in the United States, 1991–96

| Calendar year | HFCS | Corn Sweeteners | | | Total Caloric Sweeteners | Corn |
		Glucose Syrup	Dextrose	Total	Total	Sweeteners
		(Pounds, dry basis)				*(Percent)*
1991	50.0	18.2	3.8	72.0	138.0	52.1
1992	51.6	19.0	3.8	74.4	141.4	52.6
1993	54.4	19.6	3.8	77.8	144.6	53.8
1994	56.4	15.1	3.9	80.3	147.6	54.4
1995	58.2	20.6	4.0	77.3	144.9	53.3
1996	59.4	20.8	4.0	84.0	152.3	55.2

Source: U.S. Department of Agriculture, Economic Research Service, *Sugar and Sweetener Situation and Outlook Yearbook*, SSSV22N4 (December 1997).

automotive fuel. As a liquid fuel that could be used to help power a large motorized transportation sector, alcohol became a very desirable form of energy. Concurrently, the 1977 Clean Air Act and the phaseout of lead from gasoline provided added stimulus. Federal and state legislative changes were enacted and subsidies were given to encourage ethanol production for gasohol, which is one part ethanol and nine parts gasoline. Further stimulus came from the 1990 Clean Air Act Amendments to reformulate gasoline to meet certain oxygen levels to help control carbon monoxide and ground-level ozone problems (Lee, 1993).

Renewed demand and research emphasis given to ethanol manufacture, which is made primarily from corn, led to increased technical efficiency in ethanol production. As a result, corn-ethanol production has gone from a process that required 16 percent more energy than it produced to a net surplus of 33 percent. Research undertaken to lower production costs further include improvements in membrane technology, bacterial fermentation, coproduct development, and reducing costs of producing short-rotation woody crops and grass to be used as feedstocks.

Table 9.5 • U.S. Domestic Food and Beverage Uses of HFCS-42 and HFCS-55, 1980 and 1995.

Domestic Use	HFCS-42		HFCS-55	
	1980	1995	1980	1995
	1000 short tons, dry weight			
Cereal and bakery products	304	441	4	14
Confectionary, including chewing gum	11	44	1	12
Processed foods	334	657	17	163
Dairy Products	110	210	5	27
Multiple and miscellaneous	380	427	38	110
Beverages, mainly soft drinks	372	1310	525	4327
Total	1511	3089	591	4653

Source: Buzzanell, P., U.S. Department of Agriculture, "Corn Sweeteners: Recent Developments and Future Prospects," paper presented at "Azucar '95 Foro Internacional," Guadalajara, Mexico, October 10, 1995 and *Sugar and Sweetener Situation and Outlook Yearbook* SSSV21N4, December, 1996.

Table 9.6 • World Production of High Fructose Corn Syrup for Selected Countries, Selected Years.

Country	1982	1987	1992	1995
	1,000 metric tons, dry basis			
United States	2846	5145	6236	7121
Canada	110	202	250	255
Argentina	40	169	180	220
EU	260	265	286	303
Japan	579	724	761	750
South Korea	69	182	263	250
Taiwan	NA	15	125	180
Others	60	81	133	335
World Total	3964	6783	8134	9414

Source: U.S. Department of Agriculture, Economic Research Service, *Sugar and Sweetener Situation and Outlook Yearbook*, SSSV19N2 (1994) and SSSV20N4 (1995).

Technical Progress

Corn refining is a complex and highly technical process with large economies of scale requiring substantial capital investment. Also, technological improvements in recent years have reduced greatly the amount of labor needed, so that the number of workers employed decreased, from 12.1 thousand in 1972 to 9.2 thousand in 1992 even though industry capacity approximately tripled. Payroll expenses in wet corn milling declined from 42 percent of value added by manufacture in 1972 to 11 percent in 1992. As a consequence of technological advances and plant scale economies, the industry is dominated by relatively few large plants. In 1992, 26 establishments with 100 or more employees accounted for 96.8 percent of value added by manufacture. The other 25 plants produced the remaining 3.2 percent. Only a few firms account for the bulk of industry output.

Plant Location

Corn refining plants tend to be located near sources of raw material. In 1992 nearly three-fourths of industry value of shipments were from plants located in Illinois, Indiana, and Iowa (*U.S. Census of Manufactures*, 1992). These three states in the early 1990s accounted for about 45 percent of U.S. corn production. The plants in the corn belt are substantially larger than those in locations away from major corn producing areas.

Wet Corn Milling Industry Multipliers

Of economic relevance as a particular industry adds value through its ongoing operations is the income it generates for people within the industry and to others who in various ways supply that industry. Measures of the collective effects of the activities associated with value-added operations are called direct and indirect multipliers.

Based on data and a regional input-output model (IMPLAN) developed by the U.S. Forest Service, national income multipliers were estimated by Professor Kevin McNamara, Purdue University, for U.S. food manufacturing industries. The combined multiplier for the direct and indirect activities of wet corn milling in 1990 was estimated to be 3.3. That is, for each dollar of income paid by the wet corn milling sector (the direct effect), there was an additional $2.30 in income paid in the U.S.

economy by firms supporting (supplying) wet corn milling (the indirect effect).

The income dollars from direct and indirect activity are then spent and respent by households. This added household spending is called an induced impact. For wet corn milling the added multiplier effect from induced spending in the national economy in 1990 was 2.6. When the induced influence was added to the direct and indirect multiplier of 3.3, the total income multiplier became 5.9. For comparison, the average direct and indirect multiplier for food processing industries was 4.2 and the total income multiplier was 9.2. Variations occur among industries due to differences in linkages with other industries.

Also, for any given region, the impact of a particular industry multiplier varies with the size and structure of the economy of the region. If the linkages of a specific sector to other industries in the region are strong, the income multiplier of the sector will be higher than in a region where an industry is without strong linkages. Linkages to industries outside the local region (for example, Indiana firms purchasing supplies from Ohio suppliers) create leakages that take away from the state or region in which the original activity occurs. The 5.9 multiplier for wet corn milling, which was for the United States, is much larger than the 2.0 multiplier for Indiana because there were leakages from wet corn milling in Indiana to other states.

It should be realized, however, that multipliers are fully effective only if there are unemployed resources to be put to work. If resources are bid away from other industries or areas, only the net gain from their shift to wet corn milling would be the base from which the multiplier can be calculated.

Industry Organization

Although market growth and technological progress have advanced remarkably in wet corn milling in recent decades, the industry has continued to be dominated by relatively few firms. Some 70 years ago the industry was judged to be monopolistic and the leading firm, Corn Products Refining Company, was required to divest a portion of its assets. Nevertheless, this company remained the dominant firm for many years although its 60 percent market share in 1918 gradually declined to around 45 percent in 1945 (Watkins, 1927; Whitney, 1958).

In 1947, the four largest firms in Census Industry 2046, wet corn milling, accounted for 77 percent of industry value of shipments. Four-firm concentration declined to 63 percent in 1972 but rose to 74 percent in 1982, remained at that level in 1987 and was 73 percent in 1992 (Table 9.7). Only a dozen or so companies accounted for essentially all industry output.

Data are available by company for industry capacity of two important sweetener products, HFCS-42 and HFCS-55 (Table 9.8). These data show that the largest firm is Archer Daniels Midland with about one-third of total HFCS industry capacity. When capacities for the next three, A. E. Staley, Cargill, and CPC International, are included, the largest four account for around 85 percent of total capacity for manufacturing HFCS. In the early 1990s Archer Daniels Midland produced an estimated half of U.S. fuel alcohol.

Corn refining firms have become increasingly diversified, with expansion of new food products for the consumer market. They have also acquired other lines of business and become more conglomerate in character, along with their expansion through direct investment in other countries. Marion and Kim (1991) estimated that about half of the change in four-firm concentration between 1977 and 1988 came from internal growth and the other half from mergers and acquisitions. In 1991, the total sales of the four largest sweetener producing firms ranked among the 50 largest U.S. corporations (U.S. Department of Agriculture, 1994).

Conduct and Performance of the Wet Corn Milling Industry

The wet corn milling industry has been among the most concentrated in the grain and oilseeds complex. Only the breakfast cereals industry, with four-firm concentration at 85 percent of industry value of shipments in 1992, was higher than the 73 percent in wet corn milling. Because of the particular characteristics of the wet corn milling industry it is relevant to examine the industry's structure in terms of implications arising from its conceptual market form.

Wet corn milling clearly possesses the oligopolistic characteristics of few firms and, because of large economies of scale, entry by new firms is very difficult and has rarely occurred. The leading firms in the industry obviously recognize their market interdependence. A market action by one can stimulate a response by one or more others.

Table 9.7 • Percentage of Total Value of Shipments Accounted for by Largest Companies in the Corn Wet Milling Industry (SIC 2046), Selected Years.

Year	Number of companies	Total value of shipments, millions of $	Four largest companies %	Eight largest companies %	Twenty largest companies %	Primary product specialization %
1947	47	460.0	77	95	99+	88
1954	54	458.4	75	93	99	91
1958	53	522.0	73	92	99	91
1963	49	622.4	71	93	99	83
1967	32	751.3	67	89	99+	84
1972	26	832.3	63	86	99+	97
1977	22	2014.8	63	89	99+	93
1982	25	3268.4	74	94	99+	92
1987	32	4788.9	74	94	100	94
1992	28	7045.2	73	93	99+	88

Source: U.S. Department of Commerce, Bureau of the Census, *Concentration Ratios in Manufacturing* (indicated years).

Table 9.8 • U.S. Production Capacity for High Fructose Corn Syrup, by Company, 1987 and 1992.

Company	HFCS-42		HFCS-55		Total	
	1000 short tons, dry weight	Percent	1000 short tons, dry weight	Percent	1000 short tons, dry weight	Percent
1987						
Archer Daniels Midland	675	36.2	1200	31.1	1875	32.8
American Fructose	128	6.9	348	9.0	476	8.3
Cargill	355	19.0	540	14.0	895	15.7
CPC International	213	11.4	348	9.0	561	9.8
Golden Technologies	18	1.0	38	1.0	56	1.0
Hubinger	120	6.4	130	3.4	250	4.4
A.E. Staley	355	19.0	1250	32.4	1605	28.1
Total	1864	100.0	3854	100.0	5718	100.0
1992						
Archer Daniels Midland	816	30.6	1425	33.1	2241	32.2
American Fructose	284	10.7	350	8.1	634	9.1
Cargill	570	21.4	755	17.6	1325	19.0
CPC International	290	10.9	365	8.5	655	9.4
Golden Technologies	38	1.4	54	1.3	92	1.3
Hubinger	130	4.9	177	4.1	307	4.4
A. E. Staley	535	20.1	1175	27.3	1710	24.6
Total	2663	100.0	4301	100.0	6964	100.0

Source: Gray, F., P. Buzzanell; and W. Moore (1993). U.S. Corn Sweetener Statistical Compendium, U.S. Department of Agriculture, Statistical Bulletin No. 868, p. 22.

For insight into ways in which interdependent actions and reactions of firms bring about differing results, an early theoretical example developed by Cournot (Fellner, 1949) is informative. In Cournot's duopoly example, equilibrium is found at a level of price and output between pure competition and pure monopoly. Development of game theory by Von Neumann and Morgenstern (1944) provided the basis for succeeding examination of interactive conduct among business firms. However, because the range of conduct options that can be employed by some business firms is so broad, and variations in assumptions so many, economic outcomes can be very numerous and highly uncertain. Of critical importance is whether there may be strong incentives toward collaborative behavior, or collusion, that are temptingly irresistible. Although the structure of oligopoly is widespread throughout the industrial economy, the motivation toward spontaneous coordination of market actions by business firms, or the formation of coalitions or cartels, may be found primarily within the more concentrated structures. By contrast, many oligopoly structures, especially where the numbers of sizable firms are numerous, are in reality highly competitive. Thus, even within some concentrated structures there can be, in some situations, rigorous interfirm market rivalry. Consequently, empirical analyses that take into account structural aspects, along with external conditions affecting the industry and managerial characteristics within firms, are necessary to evaluate market conduct and performance.

Because some major products of the corn refining industry allow little or no opportunity for product differentiation (that is, these products of different firms in the industry are produced with strictly defined specifications and are essentially identical), the market emphasis on pricing behavior within and among firms can be critically important. There is little opportunity for use of product differentiation as a competitive strategy. Public concerns about collaborative efforts, market control, and price-fixing have thus emerged during the history of the industry, most recently in the early 1990s for lysine, an essential animo acid in animal feed (*Wall Street Journal*, 28 July and Connor, 1997). Archer Daniels Midland and four Asian companies, two in Japan and two in South Korea, pleaded guilty to price-fixing in 1996 and were assessed heavy penalties.

Price fixing is an important aspect from the standpoint of public policy, highlighting the relevance of knowledge about industry structure and conduct developments in evaluating industry performance. Of longer-term relevance is the need for information about competitive processes and industry behavior that might affect future public policy formation and implementation.

Analysis of structure and performance in corn refining is further complicated by the fact that the largest firm, Archer Daniels Midland, with its 35 percent market share in corn refining, also has 28 percent of oilseed processing and 26 percent of wheat milling. Another major firm, Cargill, has 21 percent of corn refining and about one-fourth each of oilseed processing and wheat milling (*Wall Street Journal*, 27 October 1995). Other large grain or oilseed processing firms also have sizable operations in more than one industry.

This conglomerate structure adds to complexities in analyzing and evaluating competitive behavior and performance not only of the corn refining industry but also of possible effects of interactions of corn refining firms with firms in other industries. If, for example, large conglomerate firms confront each other in several industries and markets, they may acquire tendencies toward mutual forbearance or "live and let live" competitive behavior. Additionally, firms may be able to influence government rules and policies to enhance their own competitive advantages. Thus, not only are analytical problems compounded; so are those in formulating and applying appropriate public policies that relate to the industries involved.

Effects of Corn Price Variability

Corn is the major cost ingredient used in producing corn refinery products, amounting to 81 percent of total materials, ingredients, containers, and supplies purchased by the industry in 1992 (Table 9.9). However, because the price of corn is more variable than that of other components, due primarily to corn production variability, the relative cost of corn varies considerably among years. The net cost of corn to U.S. millers, after allowance for by-product credit, was estimated at 92 cents per bushel in 1992, 26 cents in 1987, $1.03 in 1982, and 73 cents in 1977 (Gray et. al., 1993). The 56 pounds in a bushel of corn yields 31.5 pounds, dry weight, of starch, 1.55 pounds of corn oil, 2.65 pounds of corn gluten meal, and 13.5 pounds of corn gluten feed. The 6.8 pounds residual is mainly moisture. By-product credit values are based on the amounts of corn oil and corn gluten feeds times their prices.

To illustrate further, although corn accounted for 81 percent of industry costs of materials, ingredients, containers, and supplies in 1992,

Table 9.9 • Corn Price and Cost to the Corn Wet Milling Industry, 1982, 1987, and 1992.

Year	Corn price	By-product allowance	Net corn cost	Cost of materials ingredients, containers, and supplies	Corn cost	Corn cost as percent of materials, ingredients, containers and supplies	Corn cost as percent of value of product shipments
	dollars per bushel			*million dollars*		*percent*	
1982	2.48	1.45	1.03	1733.1	1370.8	79.1	44.1
1987	1.59	1.33	.26	2169.5	1435.9	66.2	32.3
1992	2.33	1.41	.92	3184.0	2587.7	81.3	40.3

Source: Gray, F., P. Buzzanel; and W. Moore (1993). *U.S. Corn Sweetener Statistical Compendium*, U.S. Department of Agriculture, Statistical Bulletin No. 868, p. 25, and U.S. Department of Commerce, Bureau of the Census, *Census of Manufacturers* (indicated years).

the percent was only 66 percent in 1987 when corn supply was abundant and prices low. These percentages were 79 in 1982 and 77 in 1977. Corn cost also varies as a component of the wet corn milling industry value of product shipments, amounting to 37 percent in 1992 and 32 percent in 1987. Because product selling prices fluctuate through a much narrower range than purchase prices for raw grain, industry earnings from corn refining tend to be inversely related to the price of corn.

International Involvement

Exports of a variety of products manufactured by the corn refining industry expanded significantly during the 1980s and 1990s, building on the continuing large foreign sales of corn gluten feed and meal. Additionally, corn oil, starches, and sweeteners all posted major gains. Total exports of processed corn milling food products reached $1.375 billion in 1992 (U.S. Department of Agriculture, 1994), which amounted to 19.5 percent of the total industry value of shipments. Some trade has occurred in high fructose corn syrup (Table 9.10). This has been primarily between the United States and Canada.

Expanded U.S. production and consumption of two major corn refinery products, high fructose corn syrup and fuel alcohol, probably contributed positively to the U.S. trade balance by substituting to some extent for imported sugar and oil. U.S. corn refining firms have also expanded their direct investment in foreign facilities. Such investment is occurring in several countries.

Future Industry Prospects

The dramatic growth in HFCS production and consumption that began in the early 1970s moderated in the late 1980s. Per capita consumption continued to rise slightly through the early 1990s. Additional HFCS marketing expansion is expected as the U.S. population increases. Also, some further penetration appears probable in selected food uses. Much will depend, however, on sugar import policies and on technological developments that might further lower manufacturing costs or lead to advances in product development, such as crystalline fructose.

Fuel alcohol production could well expand further if government policies provide incentives for more widespread use as an automotive fuel and if technical developments continue to lower alcohol manufactur-

Table 9.10 • U.S. High Fructose Corn Syrup (HFCS) Supply and Use, 1990-96

| Calendar Year | Supply | | | Utilization | | |
	Domestic Production	Imports	Total	Exports	Non Food Use	Domestic Food Use
			1000 short tons, dry weight			
1990	6258	178	6436	138	68	6229
1991	6451	159	6610	140	68	6403
1992	6656	193	6849	107	62	6680
1993	7121	189	7310	114	68	7128
1994	751C	137	7647	125	67	7455
1995	787⁴	79	7953	111	76	7767
1996[1]	8167	111	8278	203	72	8003

Source:United States Department of Agriculture, Economic Research Service, *Sugar and Sweetener Situation and Outlook Yearbook*, SSS-222, December 1997.

[1]Preliminary

ing costs. Some efficiencies have come in overall plant operating costs due to seasonal balancing of production of HFCS and fuel alcohol. Plant capacity used in producing HFCS for surging summer soft drink demands can be used during other times to produce fuel alcohol.

Food demands for other corn refinery products are expected to expand with population growth or possibly more rapidly with new product development and technological improvements that would lower unit costs. Similarly, expanded industrial uses can be expected with growth in the U.S. economy and possible new uses and applications made feasible through research.

A continued upward trend in sales abroad in both food and industrial markets appears promising. Several firms are expanding their foreign direct investments in corn refining operations and participating in joint ventures with foreign firms.

The total number of corn refining firms in the United States is not expected to change greatly, although further investment in corn refining capacity can be expected with increased demand for starch for food and industrial uses. Given the large economies of scale and highly technical nature of the business, capacity expansion is expected to continue to occur primarily within existing firms rather than through new entrants. The conglomerate character of firms with wet corn milling operations is expected to become more prevalent, although four-firm concentration may not change greatly.

References

Buzzanell, P., (1995) U.S. Department of Agriculture, "Corn Sweeteners: Recent Developments and Future Prospects," paper presented at "Azucar '95 Foro Internacional," Guadalajara, Mexico, October 10, 1995.

Connor, John M., "The Global Lysine Price - Fixing Conspiracy of 1992–95," *Review of Agricultural Economics*, Vol. 19, No.2, Fall/Winter, 1997.

Fellner, W. J., *Competition Among the Few.* Alfred A. Knopf, New York, 1949.

Gray, F., P. Buzzanell, and W. Moore, *U.S. Corn Sweetener Statistical Compendium*, U.S. Department of Agriculture, Statistical Bulletin Number 868, 1993.

Lee, H., Ethanol's Evolving Role in the U.S. Automobile Fuel Market, *Industrial Uses of Agricultural Materials,* U.S. Department of Agriculture, Economic Research Service, 1US-1, 1993.

Marion, B. W. and D. Kim, "Concentration Change in Selected Food Manufacturing Industries: The Influence of Mergers vs. Internal Growth," *Agribusiness,* Vol. 7, No. 5, September, 1991.

U.S. Department of Agriculture, Economic Research Service, *Feed Situation and Outlook Yearbook*, FDS-1995, November, 1995.

U.S. Department of Agriculture, Economic Research Service. *Food Marketing Review*, 1992-93. Agricultural Economic Report No. 678, 1994.

U.S. Department of Agriculture, Economic Research Service, *Sugar and Sweetener Situation and Outlook Yearbook*, SSSV18N2, June 1993.

U.S. Department of Agriculture, Economic Research Service, *Sugar and Sweetener Situation and Outlook Yearbook*, SSSV19N2, June, 1994.

U.S. Department of Agriculture, Economic Research Services, *Sugar and Sweetener Situation and Outlook Yearbook*, SSSV20N4, December, 1995.

U.S. Department of Agriculture, Economic Research Service, *Sugar and Sweetener Situation and Outlook Yearbook*, SSSV21N4, December, 1996.

U.S. Department of Agriculture, Economic Research Service, *Sugar and Sweetener Situation and Outlook Yearbook*, SSS-222, December, 1997.

U.S. Department of Commerce, Bureau of the Census, *Census of Manufactures*, for indicated years.

U.S. Department of Commerce, Bureau of the Census, *Concentration Ratios in Manufacturing*, for indicated years.

U.S. Department of Labor, *Monthly Labor Review*, various issues.

Von Neumann, J. and O. Morgenstern, *Theory of Games and Economic Behavior*, Princeton University Press, Princeton, NJ, 1944.

Watkins, M.W., *Industrial Combinations and Public Policy*, Cambridge, England; Houghton-Mifflin, 1927.

Whitney, S. N., *Antitrust Policies, Vol. II*, New York, The Twentieth Century Fund, 1958.

Wall Street Journal, "Investigators Suspect a Global Conspiracy in Archer-Daniels Case," Dow Jones & Co., July 28, 1995, p.1.

Wall Street Journal, "How Dwayne Andreas Rules Archer-Daniels By Hedging His Bets," Dow Jones & Co., October 27, 1995, p.1.

10

U. S. Soybean Processing Industry: Structure and Ownership Changes

Donald W. Larson

Introduction

U.S. soybean processing and grain industry structure and ownership have changed dramatically through time (Farris, 1973 and 1988; Dahl, 1992). Important among the factors affecting the ownership and structure are fluctuations in supplies of soybeans, competition from other producing countries, and changes in domestic and foreign demand for soybeans and soy products. Government policies and programs directed toward agribusinesses, farmers, and exports also affect the structure of the U.S. processing industry. The number and size of firms, ownership of firms due to mergers and consolidations, utilization of plant capacity, and crush margins have changed dramatically. The present chapter aims to: (1) describe the changing structure and ownership of the soybean processing industry; and (2) analyze the impact of some of these changes on the possible future performance of the soybean processing industry.

The United States is the world's largest producer of soybeans (about 50 percent of world production), with an average annual production of over 50 million tons (over 2 billion bushels) in recent years. The rapid growth in U.S. soybean production and processing occurred largely because of rising world demand for the beans and the primary products of oil and meal obtained from the crushing process. This rapid growth in demand derives mainly from an increased world population, rising incomes, and a desire to improve food diets, especially in the higher-income countries of Asia, European Union (EU) and North America.

Over 60 percent of U.S. soybean production is crushed each year. Soybean crush has increased steadily as production has increased, with the exception of the early 1980s when production and crush declined temporarily. For example, in 1992, from a production of 2.2 billion bushels of soybeans, 1.3 billion bushels were crushed for oil and meal, and 770 million bushels were exported. Of the domestic crush, nearly 75 percent of the soymeal is sold in the domestic market, while the remainder is exported. On the soyoil side, between 80 to 90 percent is used in the United States.

Generally, world trade in oilseeds and oilseed products has grown significantly in the 1980s; however, soybeans have lost some of their market share in the world oilseed trade to rapeseed, cottonseed and peanuts (Houston, 1992). Furthermore, U.S. export market share (about 70 percent for beans, 15 percent for soyoil, and 20 percent for soymeal) has declined from over 90 percent in 1970, due to increased competition from Argentina and Brazil (Larson and Rask, 1992). However, soybean and soybean product exports are expected to increase rapidly in the future as a result of income growth per capita and food consumption changes. Imports from Japan, Taiwan, South Korea, Mexico, the EU, and other countries are expected to increase rapidly as incomes grow and diets change to consumption of high value products.

Number and Size of Processing Plants

As a result of the worldwide production increases, soybean processing capacity has expanded in the United States and other countries, especially Argentina and Brazil. The number of plants and plant capacity in the rest of the world has increased, but in the United States the number of plants has decreased and the size of plants has increased in the last 45 years. The desire of governments to increase production and value added within their countries has created a far more competitive oilseed processing industry worldwide.

Within the United States, soybean processing plants are located primarily in the Cornbelt where soybean production is heaviest (Figure 10.1). Illinois and Iowa have the largest number of plants. In 1992, processing plants in Illinois, Indiana, and Iowa accounted for nearly 50 percent of the value of shipments of the soybean industry (U.S. Department of Commerce, 1992). According to the Soya Bluebook, the United States had 36 soybean processing firms that owned 91 plants in 1994 (Table 10.1). The Archer Daniels Midland Company (ADM) and Cargill, Inc. own 19 and 15 plants, respectively. Ag Processing, Inc. (AGP) and Bunge Corporation are much smaller at nine and eight plants, respectively. Most of the other firms own only one plant. Of the four largest firms, three are investor owned and one (AGP) is a regional cooperative owned by many farmer-owned local cooperatives.

The concentration of ownership has increased steadily through time as indicated by the fact that the four largest firms currently own 51 of the 91 processing plants (56 percent) compared to only 14 plants owned out of a total of 118 in 1946 (12 percent) and 43 plants owned out of a total of 96 in 1984 (45 percent). Because the four largest firms also own the largest processing plants, the four-firm concentration ratio of total processing capacity is even higher at 76 percent. An estimate of the four-firm concentration ratio of processing capacity increases from 46.4 in 1977 to 83.0 in 1995 (Marion and Kim, 1991; *Wall Street Journal*, 1995). The rank and share of capacity of the four largest processors in 1995 was ADM (28 percent), Cargill (25 percent), Bunge (16 percent), and AGP (12 percent) (Table 10.2).

Estimated daily industry capacity has increased steadily and has reached 121,860 tons (about 1.5 billion bushels annually) in 1992 (Table 10.3). The number of plants has declined from 118 to 99 from 1946 to 1992, and the number of firms has declined from 89 to 42 (Table 10.3). Average daily plant capacity has increased rapidly from 132 tons in 1946 to 1,231 tons in 1992, and average daily firm capacity has increased from 175 to 2,901 tons in this same period. The ratio of average daily firm capacity to average daily plant capacity has increased from 1.33 in 1946 to 2.36 in 1992. This rapid growth in firm capacity relative to plant capacity indicates a substantial ownership change accruing from larger plant

Figure 10.1 • U.S. Soybean Processing Plants, 1994

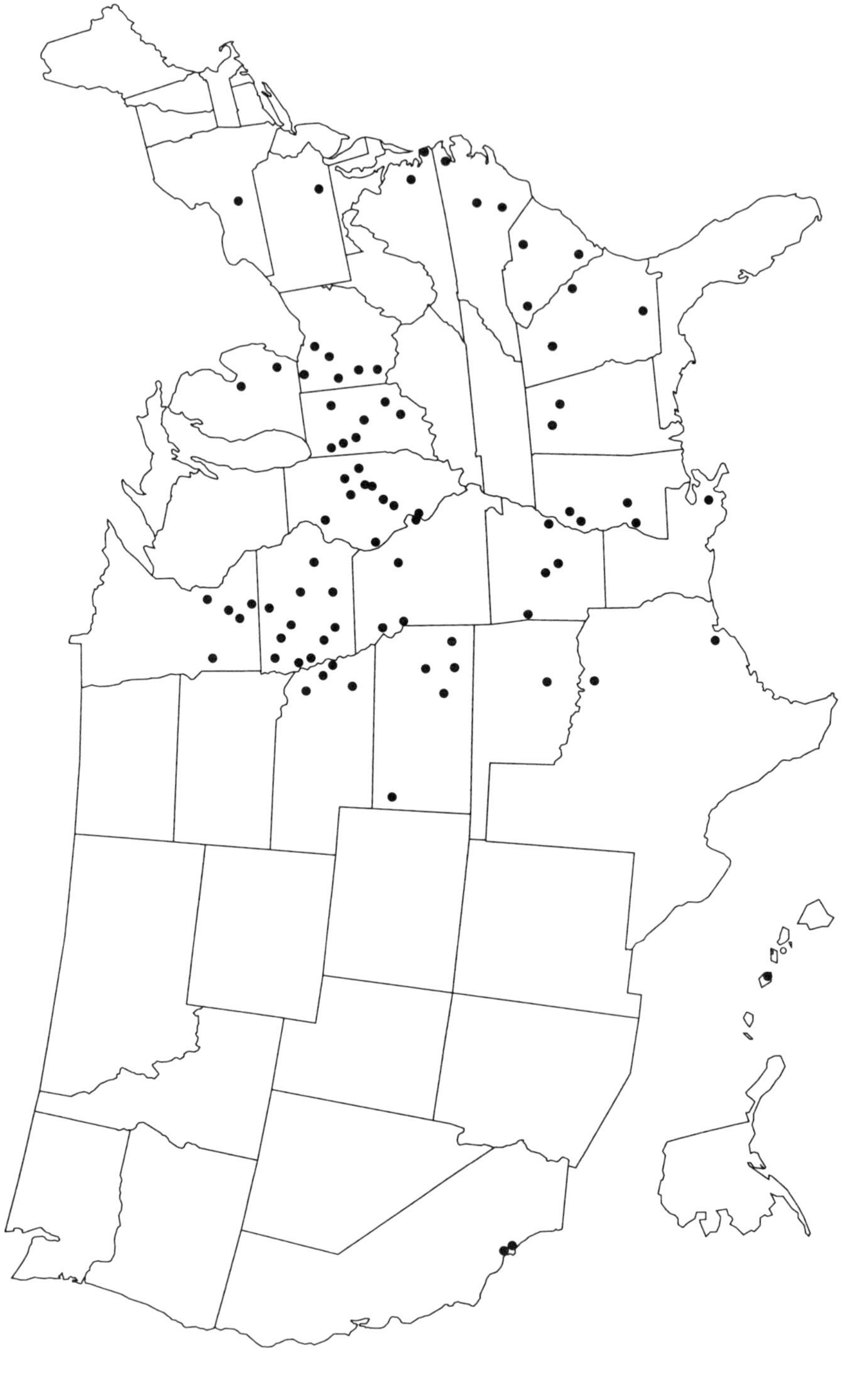

Source: American Soybean Association, *Soya Blue Book* (1994).

Table 10.1 • Soybean Processors in the U.S. and Number of Plants, 1994

Name of Firm	Number of Plants
1. Archer Daniels Midland Company	19
2. Cargill, Inc.	15
3. Ag Processing, Inc.	9
4. Bunge Corporation	8
5. Central Soya Company, Inc.	6
6 National Sun Industries	2
7. Perdue Farms, Inc.	2
8. Quincy Soybean Company[a]	2
9. Agronoico, Inc.	1
10. American Hawaiian Soy Company	1
11. American Milling Company	1
12. Bar N.A., Inc.	1
13. Central Valley Vegetable Oils	1
14. Continental Grain	1
15. Country Grown Foods	1
16. Fibred, Inc.	1
17. Grain States Soya, Inc.	1
18. Homer Oil Co., Inc.	1
19. Honeymead Products Company[b]	1
20. Houston Calco, Inc.	1
21. Lauhoff Grain Company[c]	1
22. Metamora Elevator Company	1
23. Minnesota Waxy	1
24. Nutri Meal Products Company	1
25. Pendleton Oil Mill	1
26. Producers Coop. Assoc.	1
27. Riceland Foods	1
28. Rickel, Inc.	1
29. Rose Acre Farms, Inc.	1
30. Seymour Organic Foods	1
31. Schnupp's Grain Roasting, Inc.	1
32. Southern Proteins, Inc.	1
33. Townsends, Inc.	1
34. West Bend Elevator Company	1
35. West Central Cooperative	1
36. Wysong Corporation	1
TOTAL	91

[a] Affiliate of Moorman Manufacturing Company and the C. F. Sauer Company
[b] Affiliate of Harvest States Cooperative
[c] Affiliate of Bunge Corporation
Source: *Soya Bluebook* 1994.

Table 10.2 • Top Four Firms and Four-Firm Concentration Ratio of Soybean Crushing Industry, With and Without Acquisitions 1977, 1982, 1988 and 1995

		1977		1982		1988		1995	
					Percent				
Rank	1	Cargill	15.1[a]	ADM	17.8	ADM	29.6	ADM	28
	2	ADM	14.6	Cargill	15.8	Cargill	21.9	Cargill	25
	3	Central Soya	9.9	Central Soya	9.7	Bunge	12.8	Bunge	16
	4	Ralston Purina	6.8	Bunge	7.6	Ag Processing	12.1	Ag Processing	14
Actual CR4[b]			46.4		50.9		76.4		83.0
CR4 without Acquisitions[c]			46.4		46.8		49.8		NA

[a] Percentages of total industry capacity based upon data from the directory of soybean meal and oil processors in the *Soya Bluebook*.
[b] Four-firm concentration ratio constructed based upon the actual capacity in each year of the top four firms and the total industry.
[c] Four-firm concentration ratio constructed based upon the capacity held by the top four firms in each year from originally owned or newly built plants.

Sources: Marion and Kim (1991) and *Wall Street Journal* (1995).

Table 10.3 • Estimated U.S. Soybean Processing Capacity, Number of Plants and Number of Firms, Selected Years

Year beginning October 1	Estimated daily industry capacity (tons)	Number of plants	Average daily plant capacity (tons)	Number of firms	Average daily firm capacity (tons)	Ratio of ave. firm to ave. plant cap.
1946	15,564	118	132	89	175	1.33
1953	24,829	113	220	85	292	1.33
1957	31,802	93	342	68	468	1.37
1961	41,965	97	433	65	646	1.49
1965	51,259	90	570	49	1,046	1.84
1971	79,653	87	916	43	1,852	2.02
1984	127,190	96	1,325	34	3,741	2.82
1992	121,860	99	1,231	42	2,901	2.36

Sources: Farris (1973), *Soya Bluebook*, and *Census of Manufactures*, (selected years).

and firm sizes that is increasing concentration in the soybean processing industry.

Several major grain firms expanded plant operations in the 1980s through acquisitions of existing plants (Marion and Kim, 1991, Kimle and Hayenga, 1991, and American Soybean Association, 1992, 1995). The largest crusher, ADM, acquired three plants (3,140 tons) from Buckeye Cellulose and one plant (900 tons) from Planters Manufacturing in 1982; one plant (2,400 tons) from Continental Grain and five plants (6,975 tons) from Staley in 1985 and one plant (2,000 tons) from Goldkist in 1987. ADM's daily crushing capacity increased from 18,590 in 1977 to 32,035 in 1988 as a result of these acquisitions. Cargill, the second largest crusher, purchased five Ralston Purina plants (6,300 tons) in 1985, one plant (1,500 tons) from Anderson Clayton in 1983, and one from Hartville Oil (300 tons) in 1982. The daily crushing capacity of Cargill increased from 19,190 tons in 1977 to 23,700 tons in 1988. AGP is the only cooperative among the major processors. AGP, formed in 1983, acquired three plants (4,400 tons) from Farmland, two plants of 2,500 tons from Land O'Lakes, one plant of 4,200 tons from Boone Valley Co-op, and two plants of 2,000 tons from American Grain in 1986. AGP increased its crushing capacity to 13,100 tons in 1988. Bunge acquired two plants of 2,800 tons from Cook Industry, Inc. in 1978, one plant of 2,500 tons from Goldkist in 1983, and two plants of 1,800 tons from Anderson Clayton in 1984. Bunge increased its capacity from 6,550 tons in 1977 to 13,850 tons in 1988. Cargill, Bunge, and Ag. Processing, Inc. have all increased their shares of capacity since 1988 while the ADM share has decreased slightly.

In some cases firm ownership changed without changing the number of plants. Central Soya was bought and sold by Walt Disney World and later bought by Ferruzzi (Italian company) without a change in the number of plants. Other firms such as A. E. Staley and Ralston Purina left the industry when they sold their soybean processing operations in the mid-1980s.

If one uses the *Census of Manufactures*' information, the number of soybean processing firms and plants exceeds that reported in the *Soya Blue Book*. The difference is due in large part to the fact that the latter is an industry source that does not include some of the smaller processors with the older screw press technology while the census data includes all firms and plants. Using the Census data, the number of firms increased to a peak of 68 in 1963 and declined to 42 in 1992 and the number of

Table 10.4 • Number of Firms, Plants and Capital Expenditures by Soybean Mills, Selected Years 1954–1992

Year	Number of Firms	Number of Plants	New Capital Expenditures $ Million
1954	55	88	8.5
1958	65	117	14.7
1963	68	102	10.2
1967	60	102	21.2
1972	54	94	41.9
1977	65	121	72.3
1982	52	114	113.4
1987	47	106	90.7
1992	42	99	72.3

Source: *Census of Manufactures* (selected years).

plants declined from a peak of 121 in 1977 to 99 in 1992 (Table 10.4). Although the firm and plant numbers differ somewhat between Tables 10.3 and 10.4 due to different sources and different years, one observes the same declining trend in number of firms and number of plants reported from these two sources. The *Census of Manufactures* does not report soybean processing capacity information so comparisons on processing capacity are not possible.

The number of soybean establishments by number of employees declined in all size groups during the period 1977 to 1992 (Table 10.5). The number of small plants (1-19 employees) and the number of large plants (over 100 employees) declined more rapidly than the medium sized plants. Economies of size in processing may explain the decline in the number of small plants, and introduction of labor saving technological change may explain the decline in the number of plants over 100 employees. If past growth characteristics of processing firms continue as discussed above, small plants will probably continue to decline in number in the future.

The solvent extraction process and other technological advances have led to increased scale economies, greater extraction efficiency, and increased productivity in the processing industry. The less efficient screw press technology is obsolete but still used in some small processing plants.

Table 10.5 • Distribution of Soybean Oil Mills by Number of Employees

| Number of | Number of Establishments | | | |
Employees	1977	1982	1987	1992
1-19	34	30	33	29
20-99	58	59	47	51
Over 100	29	25	36	19
Total	121	114	116	99

Source: *Census of Manufactures*, (selected years).

When major plant improvements are not feasible, some firms have selected conversion of plants to the processing of other oils; the Central Soya-owned Chattanooga, Tennessee plant converted to canola oil processing. Other less efficient plants have closed.

The economic advantages to firms who decide to integrate vertically their operations of grain origination, grain processing, and livestock feeding also increases concentration. This will increase utilization of excess elevator capacity by keeping more business in-house. Firms will use more transfer pricing to move grain from one profit center to another. For example, grain exporters already buy grain at their interior elevators for shipment on their barge lines to the Gulf for loading through their export houses (Larson, Smith, and Baldwin, 1990). A recently announced partnership between Ferruzzi and Con Agra, Inc. combines the grain origination of the Con Agra, Inc.-owned Peavey elevators and barge line with the overseas demand and export facility of Ferruzzi. Cargill has also reorganized to integrate grain origination, soybean processing, and livestock feeding to handle more grain in-house. Transfer pricing will be used increasingly to move grain among profit centers within the same company. If this becomes the wave of the future, grain supply and demand will become increasingly isolated from the marketplace with thinner cash markets. Thin cash markets will be more vulnerable to price manipulation. The opportunity for anti-competitive pricing behavior to emerge among firms increases greatly. This type of behavior is currently an issue under investigation with several large grain companies.

Total soybeans crushed increased from 27.8 million tons (927 million bushels) in 1977 to 38.1 million tons (1.4 billion bushels) in 1994 (Table

10.6). The decrease in the number of processing plants from 121 in 1977 to 99 in 1992 increased the average quantity of soybeans crushed per plant to 358 thousand tons compared with 229 thousand tons in 1977. The average crush per firm grew from 428 thousand tons in 1977 to 843 thousand tons in 1992. The amount crushed per firm and per plant was even higher in 1994 because of the continuing decline in number of firms and plants.

Increased production and efficiency has resulted in growth in such indicators as value of shipments and value added by manufacture for the period 1964 to 1994 (Table 10.6). Nominal value added increased from $170.2 million to $1,375.8 million in this period. In real terms (using GNP implicit price deflator with 1987 as a base year), this amounted to an increase from $614.4 million to $1,093.3 million in 1994.

Added capital investment was not an important factor in the growth of the soybean processing industry in recent years. The large infusions of capital in the early 1980s, such as the $113 million in 1982, have slowed to $72 million per year (Table 10.4). This decline in capital expenditures may be largely related to the acquisition of small and medium firms by larger ones. Since acquisitions are not reported as new capital expenditures, the increase in the number of large-size plants and growth in production has not been matched by increases in new capital expenditures. Decline in new investments in processing capacity might also be due to a depressed agriculture and economy in the early 1980s.

Product Value and Crush Margins

The soybean processing industry converts each bushel of soybeans into meal and oil. The crush margin is the difference between the price of soybeans and the value of the soybean products: meal and oil. The margin indicates the cost, including profit, of providing crushing services and is influenced by many factors. These include the purchasing practices of the processor, location and size of the firm, fluctuations in soybean supply and demand, competition for soybean purchases, and the product yield per bushel of soybeans. The value of soybeans, in turn, depends on the prices and yields of oil and meal.

The amount and value of meal obtained from processing a bushel of soybeans exceeds that from oil. During the 1980–87 marketing years, the average yield was 47.3 pounds of meal and 11 pounds of oil per 60-pound bushel. Oil sold for 22.3 cents per pound compared to 9.03 cents for

Table 10.6 • Crush, Value of Shipments, Value Added by Manufacture, and Crushing Margins in the U.S. Soybean Processing Industry, 1964–94

Year	Crush (million bushels)	Nominal value of shipments ($ Billion)	Value added by Manufacture Nominal ($ Million)	Real[a] ($ Million)	Crush Margins Nominal ($)	Real[a] ($)
1964	479	1.59	170.2	614.4	0.26	0.94
1965	537	1.73	157.4	554.2	0.26	0.92
1966	559	2.01	209.5	712.6	0.15	0.51
1967	576	2.15	215.4	710.9	0.12	0.40
1968	606	1.92	254.1	799.1	0.11	0.35
1969	737	2.03	239.1	715.9	0.48	1.44
1970	760	2.62	410.7	1,166.8	0.26	0.74
1971	721	2.86	334.8	902.4	0.09	0.24
1972	722	3.36	350.0	899.7	0.59	1.52
1973	821	5.27	623.2	1,509.0	0.72	1.74
1974	701	7.32	691.9	1,541.0	0.16	0.36
1975	865	6.16	340.1	691.3	0.16	0.33
1976	790	6.74	624.2	1,193.5	0.19	0.36
1977	927	7.58	373.8	668.7	0.28	0.50
1978	1,018	8.23	612.9	1,016.4	0.36	0.60
1979	1,123	9.09	907.9	1,384.0	0.40	0.61
1980	1,020	9.75	846.4	1,180.5	0.33	0.32
1981	1,030	9.18	819.8	1,039.0	0.43	0.54
1982	1,108	8.60	678.2	809.3	1.05	1.25
1983	983	9.06	888.1	1,018.5	0.30	0.34
1984	1,030	9.99	651.6	715.3	0.45	0.49
1985	1,053	8.63	711.7	753.9	0.60	0.70
1986	1,179	7.82	676.7	698.3	0.68	0.73
1987	1,174	9.07	1,011.5	1,011.5	1.42	1.31
1988[b]	1,058	12.14	1,411.4	1,358.4	0.89	0.86
1989[b]	1,146	10.72	1,231.4	1,130.8	1.13	1.04
1990[b]	1,187	10.97	1,519.0	1,341.9	0.77	0.68
1991[b]	1,254	9.97	1,186.5	1,008.1	0.91	0.77
1992[b]	1,279	10.66	1,274.8	1,054.4	0.89	0.74
1993[b]	1,276	N/A	1,330.7	1,077.5	0.96	0.78
1994[b]	1,405	N/A	1,375.4	1,093.3	1.18	0.94

[a] GNP implicit price deflator (1987=1.0) was used.

[b] Crush margin obtained from *Oil Crops Situation and Outlook*, (selected years).

Source: *Soya Bluebook* (1991) and *Oil Crops Situation and Outlook*, (selected years).

meal; therefore, meal accounted for 64 percent of the total value and oil represented 36 percent. Given the strong demand for meal by the domestic livestock sector and a world surplus of edible oils , soymeal share of total value did increase in the 1990s (Oil Crops Situation and Outlook, 1994).

Crushing margins averaged 73 cents per bushel during the 1980s, almost five times the estimated break-even level of 15 cents (Schaub et al., 1988). In 1987, the sharp price increases of both meal and oil led to the highest crushing margin ($1.42) of the period (Table 10.5). In real terms, however, the crushing margin was lower in the 1970s, averaging $0.70 per bushel, than in the 1980s. The margin increased to an average of $0.76 per bushel in the 1980s due in part to the structural reorganization and efficiency gains discussed above.

Crushing margin trends were measured, using ordinary least square regression procedure, for the period 1964–1994 (Table 10.7). Significant positive trends were found in crushed soybeans, nominal value added by manufacture, and nominal crushing margin. The trend in crushed beans was 26.76 million bushels per year. The trend in nominal value added

Table 10.7 • Soybean Crush and Crush Margin Regression Results, 1964–94

Dependent Variable[1]	Trend Coefficient[2]	Intercept	R-Squared
Million bushels crushed	26.76 (1.36)	–52,038	0.93
$ Million value added:			
Nominal	42.22 (3.19)	–82,852	0.86
Real	10.6 (5.16)	–19,950	0.13
$ Crush margin per bushel:			
Nominal	0.03 (0.005)	–61.3	0.59
Real	0.002 (0.008)	–4.14	0.003

Source: Estimated from Table 10.6.

[1] Independent variable is the four-digit year, 1964–94.

[2] Standard error of the coefficient is in parenthesis.

was $42.2 million per year, and that for the nominal crush margin was 3 cents per bushel per year. In real terms (1987 as base year) the results were mixed, especially for the crushing margin. The positive trend for value added by manufacture equalled $10.6 million per year, while the trend for real crushing margin was 0.002. The trend in the real crushing margin is positive and very small but not statistically significant, so that no definitive conclusion on the trend in real crushing margin is possible from 1964 to 1994.

Government policies toward prices, output, trade and industry structure influenced industry performance in the 1980s, sometimes in different directions. Government price support programs, for instance, could benefit soybean processors considerably by reducing downside price risk when soybean prices are near the support level (Crowder and Glauber, 1990). On the other hand, by setting a floor for domestic prices of soybeans, U.S. price support programs raise raw material costs for processors. Price support programs for other grains, especially corn, also limit expansion of soybean production and processing because farmers plant to protect their corn base rather than respond to market incentives that want more soybean production. Marketing loan programs, on the other hand, are expected to help processors and exporters better compete in export markets.

Conclusions

Many of the factors that influence structure and capacity utilization in the soybean processing industry changed in the 1980s. The supply of beans and the amount of soybeans crushed increased. Demand for soymeal, and to a lesser degree for soyoil, increased in the United States. On the world market, U.S. exports of soybeans and soy products faced stiff competition from such countries as Argentina and Brazil. Market shares for these countries increased in the 1980s while U.S. shares declined in part because of U.S. price support programs that idled land that could produce soybeans and propped up soybean prices.

Industry indicators like the four-firm concentration ratio increased during the 1970s and 1980s. The four largest firms' share of soybean processing capacity increased from 46 percent in 1977 to 83 percent in 1995. The number of firms decreased from 89 in 1946 to 42 in 1992, and the number of plants decreased from 118 to 99 in the 1946 to 1992 period. Average crushing capacity per plant and per firm has increased rapidly as a result of the changes. The number of small crushing plants (1 to 59 employees) decreased most rapidly while the number of large plants increased. Mergers, acquisitions, and consolidations explain the dramatic change in structure. Much of the structural change appears motivated by excess capacity and razor thin operating margins in grain merchandising, causing firms to search for new business opportunities in more profitable value-added businesses.

Major grain trading companies also became major soybean processors in the 1980s. In many firms, these mergers and acquisitions impacted the grain trading as companies saw opportunities to vertically integrate their operations. Typically, the large companies were already integrated financially but not operationally. Each operating division was a separate profit center that would buy and sell grain where the opportunities were best whether inside or outside the company. Firms owning elevators, processing plants, and livestock feeding operations decided to integrate operations to keep more business in-house to utilize the excess capacity. Grain origination in elevators would complement the demand at processing and feeding operations within the company. The operations would not buy and sell grain but would use transfer pricing to move grain among profit centers. These changes may produce less competitive grain markets that lead to anti-competitive behavior in the buying and selling of grain. Consumers and farmers may see the effects of these trading practices in terms of larger operating margins, lower farm prices, and higher consumer prices.

References

American Soybean Association. "Telephone interviews in September, 1992 and 1995," St. Louis, Missouri.

Crowder, Bradley M. and Joseph W. Glauber. "Government Programs for Soybeans," *National Food Review*. Volume 13. Economic Research Service. U.S. Department of Agriculture. March, 1990. pp. 32-36.

Dahl, Reyold P. "The Changing Structure of the United States Grain Marketing System," Staff Paper P92-23, Department of Agricultural and Applied Economics, University of Minnesota, St. Paul, Minnesota. October, 1992.

Farris, Paul L. "Changes in Number and Size Distribution of U.S. Soybean Processing Firms," *American Journal of Agricultural Economics*. August, 1973. pp. 495-499.

Farris, Paul L., Richard T. Crowder, Reynold P. Dahl, and Sarahelen Thompson. "Economics of Grain and Soybean Processing in the United States," Chapter 9 in *Economics of Food Processing in the United States*. Paul L. Farris, edt. New York: Academic Press, Inc. 1988. pp. 315-347.

Houston, Jack E., Christopher S. McIntosh, and Jamal Othman. "Structural Change in the U.S. Oilseeds Processing Industry," Paper presented at a North Central Grain Marketing Symposium on Performance of the U.S. Grain Marketing System Minneapolis, Minnesota October 12-13, 1992.

Kimle, Kevin L. and Marvin L. Hayenga. "The Structure of U.S. Grain Industries," Unpublished Paper, Department of Economics, Iowa State University, Ames, Iowa. Dec. 1991.

Larson, Donald W., Thomas R. Smith and E. Dean Baldwin. "Soybean Movements in the United States,1985". North Central Regional Bulletin No. 323, Southern Cooperative Series Bulletin No. 345, and University of Illinois Bulletin No. 792, University of Illinois, Champaign, Illinois. 1990, 45p.

Larson, Donald W. and Norman Rask. "Changing Competitiveness in World Soybean Markets," *Agribusiness: An International Journal.* 1992. Vol. 8. No. 1. pp. 79-91.

Marion, Bruce W. and Donghwan Kim. "Concentration Changes in Selected Food Manufacturing Industries: The Influence of Mergers vs. Internal Growth," *Agribusiness: An International Journal,* Vol. 7, No. 5, September, 1991, pp. 416-431.

Schaub, James, W.C. McArthur, Duane Hacklander, Joseph Glauber, Mack Leath and Harry Doty. "The U.S. Soybean Industry," Agricultural Economic Report Number 588. Economic Research Service. U.S. Department of Agriculture, Washington, D.C. May, 1988.

"Oil Crops Situation and Outlook," Economic Research Service, U. S. Department of Agriculture, Various Issues.

Soya Bluebook, 1989, 1991 and 1994. Soyatech, Inc. Bar Harbour, ME. 1989, 1991 and 1994.

U.S. Census Bureau. Census of Manufactures, Selected Years. Washington, D.C.

U. S. Department of Commerce. "Current Industrial Reports, Survey of Plant Capacity, Selected Years," Washington, D.C.

Wall Street Journal, "Risk Averse: How Dwayne Andreas Rules Archer-Daniels by Hedging His Bets," Wall Street Journal, New York City: New York. October 27, 1995, p. 1.

11

Structural Change in the U.S. Edible Vegetable Oils Processing Industry

Jack E. Houston

Introduction

Changing attitudes toward foods which are beneficial or harmful to health have become a major force in shaping consumption patterns in the United States. Red meats and animal fats are notable examples of changing consumption attitudes and demands. Through the 1970s, the edible fats and oils sector had been largely characterized by a high degree of commodity interchangeability due to what were perceived as minor differences in the physical and chemical characteristics of these oils (Augusto and Pollak, 1981). Commodity promotion, product advertising, and government health agency or private consumer advocate groups' notices in public media have stimulated changes in consumer attitudes in the 1980s. Changes in consumer attitudes and consumption of various fats and oils have, in turn, induced new or modified products and processes and encouraged industrial behavioral and structural changes.

At least three events have sparked public comment and significant change in the industries encompassing the oilseeds–edible oils and fats sector: the certification of edible rapeseed oil/canola development; an American Soybean Association promotion raising health issues associated with the use of tropical oils and subsequent food labeling legislation; and an emergency Chicago Board of Trade action in soybean futures. Comments on developments in the edible oils and fats industry must include these events and potential changes in structure and behavior induced by them.

The U.S. Edible Oils and Fats Refining Industry

The edible oils and fats refining industry is comprised of plants that handle one or more continuous processing operations used to refine crude fats and oils for edible end use. The U.S. refining industry became more concentrated in the period between 1975 and 1982, according to a 1982–83 marketing year USDA survey (Willimack, 1985). From 97 refining plants operated by 49 companies in 1975 (representing 100 percent of industry processing), the number of refining plants and companies for all fats and oils declined to 58 plants operated by 29 companies (representing 88 percent of industry processing) in 1982–83. Potential optimum practical refining capacity, as indicated by the 1982–83 USDA survey, was 16.9 billion pounds, up approximately 17 percent from the 1975 survey. However, capacity utilization was still a low 76 percent. The 1982–83 survey respondents attributed the limited attainment of optimum practical output to the lack of adequate markets for finished products. By 1989, however, capacity utilization in the fats and oils sector had increased to 89 percent (*Food Marketing Review*, 1991).

Nine major types of vegetable oils are traded in the U.S. domestic market—soybean, sunflower seed, rapeseed (canola), peanut, coconut, palm kernel, palm, corn, and cottonseed oils. Nearly 75 percent of the total U.S. refined output of fats and oils in 1982–83 was soybean oil, followed by corn oil at over 7 percent, and cottonseed oil at 6.4 percent. Other vegetable oils, which included palm, palm kernel, coconut, peanut, and rapeseed, accounted for 6.7 percent of the total. Of the 58 refining operating plants in 1982–83, about half had some sort of packing facility at the refinery plant, while nearly all shipped products in bulk to other plants for product manufacture and packaging. Half of the edible fats and oils manufactured was shipped to manufacturers of shortening, margarine, mayonnaise, and salad dressing. Other outlets for processed edible oils and fats include food manufacturers, confectionery use, restaurants, food service and institutional distributors, retail grocers and distributors, industrial and inedible use, government, and feed manufacturers.

Technical Environment of Edible Fats and Oils

The bulk of fats and oils production reaches the final consumer in processed form. The edible demand for fats and oils is thus a derived demand—the nature of the market depends on the level of demand for the end-use products by the final consumers. Substitution is relatively easy in the major products—baking and frying fats, salad and cooking oils, and margarine — because of the availability of several domestically-produced vegetable oils with lower saturated fat content. Although they are technically interchangeable, costs of refining, specific end use requirements, and consumer tastes and preferences limit the range within which each of the individual fats and oils actually are substituted.

All vegetable oils have saturated fatty acid contents, but the degree varies among individual oils. Among major vegetable oils, coconut oil is the most saturated at 87 percent, followed by palm (49 percent), cottonseed (26 percent), peanut (17 percent), soybean (15 percent), corn (13 percent), sunflower (10 percent), and canola (6 percent) oils. The major U.S. edible uses of about 14.2 billion pounds of fats and oils in 1987–88 were for ingredients in salad and cooking oils (6.5 billion pounds), baking and frying fats or shortening (5.4 billion pounds), and margarine (1.9 billion pounds) (USDA, *End Uses of Fats and Oils*). Soybean oil dominated, accounting for about 10.4 billion pounds, followed by corn oil (947 million pounds), cottonseed oil (800 million pounds), and peanut oil (300 million pounds). Animal fats (lard and tallow) constituted about 1.1 billion pounds, while palm and coconut oil together totaled only 430 million pounds or 3 percent of total edible use (*Agricultural Outlook*, May 1989).

Each edible oil lays claim to two distinct markets—the market in which it has a qualitative advantage over other oils for specific attributes and the market in which it competes directly with other edible fats and oils. Consumption patterns in the first market depend on the demand for the end products which require its special properties. Consumer perceptions about which oil is healthy or harmful and other consumer tastes and preferences play a major role in determining the derived demand for the respective oils. Demand in this specialty market is theoretically less elastic than in the second market (Augusto and Pollak, 1981).Half of the edible tropical oils use is for baking and frying fats. Substituting other vegetable oils for tropical oils does not necessarily eliminate saturated fats. Hydrogenation, a process that raises the melting point of oil and improves shelf life, makes it possible to replace tropical oils with other vegetable oils, but it increases saturated fat levels in those oils. The "other or specialized uses" category provides tropical oils their niche. Coconut, palm, and palm kernel oil are used in coffee whiteners, whipped toppings, confectionery products, and cracker spray coatings because they impart to food desirable texture, taste, and appearance that other oils could not duplicate (Augusto and Pollak, 1981).

International and U.S. Vegetable Oils Consumption and Trade

While the United States is the world's largest producer of soybean oil, it is also the largest single consumer. The excess must be stored or exported. All tropical oils used in the United States are imported. Malaysia produced close to 90 percent and 80 percent of U.S. imports of palm and palm kernel oil, respectively, in the years 1986 through 1989. The Philippines accounted for about 87 percent of U.S. coconut oil imports (*Agricultural Outlook,* May 1989). Canola oil "obtained a 'generally-regarded-as-safe' for use as fats and oils and food status by the U.S. Food and Drug Administration (FDA) in early 1985" (Emery, 1985, p. 26T), which paved the way for increased competition from mostly imported canola oil in U.S. domestic markets.

Worldwide, soybean oil ranks first in edible oil consumption, accounting for over 26 percent of world use in 1991, down from just over 33 percent in 1979 (Table 11.1). Among its rivals, palm oil ranked second to soybean oil in consumption but first among the internationally-traded vegetable oils. Palm oil also demonstrated the largest average annual global increases in per capita consumption in the 1970s and early 1980s, competing aggressively with soybean oil in terms of price. For example, when the average price differential between soybean and palm oil went from 4.2 cents in 1973–74 to 9.3 cents in 1974–75, U.S. palm oil imports and consumption greatly increased, reaching about 10 percent and 8

Table 11.1 • World Food Use Consumption of Major Edible Vegetable Oils, 1979–1991.

Year	Soybean	Peanut	Cottonseed	Rapeseed	Coconut	Palm
			Food Use Consumption			
			1,000 metric tons			
1979	11,793	2,884	2,971	2,984	1,530	4,062
1980	12,068	2,564	3,043	3,680	1,967	4,742
1981	12,555	3,050	3,212	4,133	1,836	5,053
1982	12,628	2,635	2,996	4,553	1,749	5,066
1983	12,707	2,886	2,878	4,522	1,497	5,088
1984	12,784	3,005	3,682	5,074	1,582	5,515
1985	13,244	2,866	3,314	5,555	1,843	6,343
1986	14,253	3,068	3,107	6,046	1,795	6,677
1987	14,533	3,162	3,482	6,713	1,647	6,973
1988	14,438	3,773	3,625	6,939	1,726	7,596
1989	15,314	3,431	3,252	7,225	1,858	8,692
1990	15,213	3,521	3,550	7,859	1,846	8,679
1991	15,336	3,548	3,664	8,141	1,875	9,343

Source: U.S.D.A., P S & D VIEW '91 (computer file)

percent of the domestic disappearance of U.S. soybean oil and total vegetable oils, respectively, from less than 1 percent of each. Growing U.S. imports of palm oil in the mid-1970s led to calls for import tariffs by U.S. vegetable oil producers and processors. U.S. imports declined in the late 1970s, but grew again in the mid-1980s (Table 11.2). U.S. soybean oil producers and processors, through the American Soybean Association (ASA), claimed that palm and coconut oil imports significantly lowered the domestic disappearance of soybean oil, in 1986 displacing soybean oil equal to 171 million bushels of soybeans (*Soybean Digest,* September 1987).

The American Soybean Association

Founded in 1920 and incorporated in 1946, the American Soybean Association is a national, non-profit, volunteer, single-commodity organization of some 26,000 soybean farmers (1986), several associate and foreign contributing members, and businesses and persons engaged in the promotion and interests of soybeans or soybean products. ASA promotes but does not sell soybeans or soybean products. The ASA's objectives include promoting profitable soybean production for U.S. farmers, monitoring and influencing U.S. governmental programs and policy to represent the best interests of U.S. soybean farmers, promoting worldwide utilization of U.S. soybeans and soybean products, assuring a free world market for soybeans, assuring adequate support for soybean research, and informing U.S. soybean farmers of current production and marketing information (*Soya Bluebook,* 1986). Farmer-governed boards allocate checkoff funds to the American Soybean Development Foundation (ASDF). The ASDF contracts with the ASA to implement market development, research, and education programs.

ASA "Tropical Oils" Campaign

Although medical journals had long cited health issues associated with various animal and vegetable oils, widespread consumer attention was drawn to "tropical oils"—palm, palm kernel, and coconut oils—in U.S. edible oil markets in 1986. In that year, the ASA launched a campaign which highlighted alleged health risks associated with the relatively high saturated fat content of tropical oils compared to soybean oil (*Agricultural Outlook,* May 1989). The National Heart Savers Association and the National Heart, Lung and Blood Institute subsequently promoted oils low in saturated fats. In early 1987, two bills (S. 1109 and H.R. 2148)

Table 11.2   **U.S. Imports of Palm and Coconut Oils, 1979–1991.**

Year	U.S. Imports		Share of U.S Consumption		Share of Imported Vegetable Oils		Food Use Proportion of Imported Oil	
	Palm	Coconut	Palm	Coconut	Palm	Coconut	Palm	Coconut
	1000 Metric Tons				*percent*			
1979	96	367	2.1	7.0	16	63	79	32
1980	147	509	2.7	9.0	19	66	88	32
1981	99	436	1.9	8.0	15	65	77	34
1982	140	401	2.3	7.1	20	57	87	39
1983	168	416	3.2	7.5	23	57	86	33
1984	169	404	2.9	6.7	21	51	85	32
1985	277	552	4.4	7.5	25	50	87	33
1986	215	493	3.5	7.5	20	46	55	30
1987	164	487	2.4	6.4	14	41	58	25
1988	135	353	2.2	6.4	13	34	65	23
1989	113	471	1.6	5.5	10	44	70	25
1990	135	420	1.8	5.6	12	38	74	24
1991	140	400	1.8	5.3	12	35	72	23

backed by the ASA were introduced into the U.S. Congress to require labels noting tropical oils are "a saturated fat" when used in food products (*Soybean Digest*, November 1987). The bills were rejected, however, because they would selectively discriminate against types of oils whose health characteristics were not yet fully understood (*Foreign Economic Trends*, June 1989).

Proponents of the tropical oils attributed the action of the ASA to the fear of losing markets. Consumption patterns for major vegetable oils in the U.S. domestic market changed considerably following commencement of the ASA campaign. While monthly per capita edible consumption for soybean and palm oil demonstrated steady gains until late 1986, soybean oil consumption then leveled and palm oil use declined dramatically. Coconut and cottonseed oil edible consumption remained somewhat stable until late 1986/early 1987, when coconut oil began a general decline. Cottonseed oil enjoyed an exceptional increase, as did rapeseed/canola, with a variety newly developed and approved for edible use in 1985. Monthly mean per capita edible consumption for cottonseed oil increased by 20 percent directly following the ASA campaign, accompanied by immediate declines of 17 percent and 16 percent for U.S. palm and coconut oil consumption, respectively. U.S. imports of tropical oils declined sharply, as did their share of total U.S. vegetable oil consumption (Table 11.2).

Promotional Impacts on U.S. Edible Vegetable Oils Consumption

The campaign strategy taken by the ASA in combating tropical oils appeared to have been effective in U.S. domestic markets. Soon after the campaign commenced, several large food manufacturers replaced tropical oils with other vegetable oils in response to growing perceptions of health concerns (*Soybean Digest*, December 1987). Other food manufacturers also promised to abandon palm oil usage by the end of 1989. Following numerous communications and meetings between members of the American Edible Oils and Fats Industry and the Malaysian Palm Oil Industry, the ASA decided to stop the "palm oil war." In March 1989, the ASA redirected its public promotion efforts to emphasize the positive attributes of soybean oil without making negative comparisons to any other edible oils.

The question posed is, what can be deduced from the empirical evidence provided on the demand structure for the edible vegetable oils in the periods prior to and following commencement of the ASA campaign?

Whether the ASA campaign was a trade or a health issue, empirical evidence supports the hypothesis that the demand for certain vegetable oils was structurally changed in the period following commencement of the ASA health issues campaign. Studies by Othman et al. (1989, 1993, 1994) developed U.S. processor level demand models for soybean, cottonseed, palm, and coconut oils to analyze whether structural changes had been effected on the edible demand for the four oils between the periods prior to and following commencement of the ASA–National Heart Savers campaigns.

The period prior to the campaign was that period where market forces (relative prices) were assumed to be dominant; i.e., choice of edible oils by food processors was not significantly influenced by consumers' health concerns regarding saturated fats. The period following commencement of the campaign was hypothesized to show changed consumers' marginal utility associated with the different vegetable oils because of the negative ASA promotion. Cottonseed oil was included because similarities in its end-use patterns made it most readily substitutable with soybean oil (Fryar and Hoskin, 1985). A by-product of high fructose corn syrup (HFCS) and alcohol production processes, and little-used in the price-sensitive shortening sector, corn oil is generally insensitive to prices in the oils complex (Fryar and Hoskin, 1985).

Derived demands for soybean, cottonseed, palm, and coconut oils were each specified as a function of own price, prices of other vegetable oils, disposable personal income, and lagged demand, allowing for a partial adjustment process in month-to-month use related to flows of stocks, contracts, consumption habits, and other factors limiting immediate adjustment. Corn and peanut oil prices were included in the system. Structural change was defined as any change in the form of final consumptive demand, as expressed by significant change in the derived demand functions for key ingredients which are differentiated by taste and preference attributes of end-use products. Plots of consumption per capita of each edible oil over time and its price differential relative to that of soybean oil supported the hypothesis of structural change (Othman et al., 1994a). The period prior to the campaign included January 1980 through August 1987, and the period following commencement of the campaign included September 1987 to December 1990.

Following commencement of the ASA campaign, edible consumption of palm and coconut oils generally demonstrated downtrends coincident with downtrends in their price differentials, contrary to economic theory. A likelihood ratio test (Judge et al., 1885, pp. 472-476) was used to test the hypothesis of significant structural changes for palm and coconut oils between periods which suggested that the ASA campaign successfully stimulated changes in the patterns of edible palm and coconut oils consumption. Industry pronouncements relating the decline in imports and edible consumption for tropical oils during that period support this inference (*Soybean Digest*, 1989).

Short-run, own-price elasticity for edible soybean oil was quite inelastic, and short-run, cross-price elasticities for soybean oil with respect to other edible oils, including palm and coconut oils, were negligible and mostly insignificant (Table 11.3). The results suggested that U.S. domestic demand for soybean oil was not generally sensitive to prices of other edible vegetable oils, particularly to the tropical oils or to peanut oil, another competitor in world markets. The significant exception was corn oil. Cottonseed oil consumption appeared similar to that for soybean oil, showing substitution cross-price relationships but a somewhat greater short-run, own-price elasticity. Although cottonseed oil exhibited less seasonality effects than soybean or tropical oils, seasonal tendencies were consistent throughout the edible vegetable oil markets.

The short-run, (month-to-month) own-price elasticity for palm oil appeared to become positive in the period immediately following commencement of the ASA campaign. That is, after the campaign, manufacturers more readily switched from palm to other edible vegetable oils, even though palm oil prices were declining relative to soybean prices, which provides further evidence that food manufacturers willingly shunned lower-priced substitutes in response to increasing consumer concerns about health issues raised in the ASA promotions. Edible palm and coconut oils show more complementary relationships to other oils than do the soft vegetable oils, suggesting more specialized uses in end products. Cross-price responses between edible consumption of palm and coconut oils were not only complementary but highly intertwined. Coconut oil exhibited a relationship similar to that of palm oil, due to its closeness with respect to some chemical properties and end uses. Coconut oil's own-price response remained quite inelastic but more than doubled in the period following the ASA tropical oils campaign, and its cross-price responses generally showed complementary relationships. This evidence supports the assertion by Nyberg (1970) that lauric oils have few close substitutes.

U.S. derived demands for major edible vegetable oils—soybean, cottonseed, palm, and coconut oils—were highly inelastic with respect to own prices and appeared to have few significant substitutes in the short

Table 11.3 • Estimated Own-Price, Cross-Price, and Income Elasticities of Soybean, Cottonseed, Palm, and Coconut Edible Oil Demand in the Ante- and Post-Campaign Period.

Elasticities	Soybean	Cottonseed	Palm	Coconut
Ante-Campaign				
Soybean	−0.217	0.277	0.103	−0.117
Cottonseed	0.127	−0.437	0.102	0.199
Palm	0.025	0.049	−0.178	−0.322
Coconut	−0.013	0.031	−0.180	−0.108
Peanut	−0.015	−0.004	0.284	0.131
Corn	0.216	0.033	−0.245	−0.148
Income	0.406	0.759	0.840	−0.022
Post-Campaign				
Soybean	−0.217	0.277	0.749	0.124
Cottonseed	0.127	−0.437	−0.918	−0.269
Palm	0.025	0.049	0.668	0.254
Coconut	−0.013	0.031	−0.611	−0.269
Peanut	−0.015	−0.004	−0.060	−0.401
Corn	0.216	0.033	−1.258	−0.196
Income	0.406	0.759	1.026	−1.321

run. Significant changes in the derived demand responses for edible palm and coconut oils accompanied a downward shift in actual consumption and market shares in the United States. The combined share of U.S. vegetable oil consumption held by palm and coconut oils declined from 11.9 percent in 1985 to 7.1 percent in 1991, while the food use proportion of palm oil consumed in the United States dropped from 87 percent in 1985 to 55 percent in 1986.

Permanent or Temporary Structural Change in U.S. Edible Vegetable Oils?

Palm and coconut oil competitiveness with soybean and other U.S. domestic oils in the 1980s was later examined using a time-varying parameter regression model (Othman et al., 1994a and 1994b). While perceived characteristics of a particular good determine the relative preference for that good, advertising and promotional activities contrive to influence consumers' preferences among goods by providing selective information about the characteristics of the good or closely-related, competing goods (Goddard and Amuah, 1989; Goddard and Glance, 1989). But the question this time was, are the influences of such promotions relatively permanent, or are they likely to be temporary deviations from longer-term trends or cycles?

The competitive effects of the ASA tropical oils campaign on edible vegetable oil markets in the United States, using a time-varying parameters model, measured the relative magnitudes of the permanent and temporary portions of the time-varying parameter vectors and changes in demand elasticities for U.S. soybean, palm, coconut, and cottonseed oils. To the degree that technical substitutability exists among vegetable oils, relative prices play an important role in determining which vegetable oils are purchased by food processors. Refining costs, specific end-use requirements, dietary concerns, influences on habit formation, and contractual obligations limit the range within which the individual oils are actually substituted, especially in the short run.

U.S. domestic derived demands for soybean, cottonseed, palm, and coconut oils again were related to own price, prices of other vegetable oils, disposable personal income, and lagged consumption, and parameters in the model were hypothesized to be subject to both permanent and transitory changes. A time-varying parameter approach, following Cooley and Prescott (1973a, 1973b), estimated the short-run (monthly) demand elasticities. Data again included monthly observations from January 1980 through December 1990, with September 1987 hypothesized to represent observed structural changes in vegetable oils consumption.

The own-price elasticity for soybean oil demand, although highly inelastic in both periods, doubled in the period following the ASA promotion, and, on average, the elasticity was more variable in the post-campaign period than in the pre-campaign period (Table 11.4). Little difference was observed in the estimated cross-price elasticities of soybean oil with respect to other domestic oils, although again the cross-price elasticities varied more in the post-promotion period. The soybean cross-price elasticity with other domestic oils remained very low, averaging 0.086 to 0.115, respectively, in each period.

Edible tropical oils demonstrated a very weak complementary relationship with soybean oil, and the relationship between soybean oil and tropical oils weakened in the post-campaign period. These findings support Reed et al. (1985) that U.S. soybean oil demand was quite insensitive

Table 11.4 • Mean Estimated Own- and Cross-Price Elasticities and γ for Soybean, Cottonseed, Palm and Coconut Oil Demand in the Ante- and Post-Campaign Periods.

Parameter	Ante-Campaign				Post-Campaign			
	Soybean	Cottonseed	Palm	Coconut	Soybean	Cottonseed	Palm	Coconut
γ	0.056 (0.036)	0.043 (0.047)	0.092 (0.148)	0.096 (0.104)	0.046 (0.036)	0.066 (0.144)	0.132 (0.226)	0.226 (0.269)
Elasticities:								
Own–Price	−0.027 (0.030)	−0.477 (0.193)	−0.161 (0.064)	−0.067 (0.067)	−0.058 (0.054)	−0.442 (0.137)	−0.225 (0.138)	−0.043 (0.074)
Cross-Price Other Domestic	0.086 (0.035)	0.394 (0.157)	−0.392 (0.115)	−0.227 (0.123)	0.115 (0.066)	0.449 (0.123)	−1.276 (0.715)	−0.819 (0.451)
Cross-Price Tropical	−0.044 (0.027)	0.032 (0.055)			−0.034 (0.030)	0.021 (0.040)		
Cross-Price Soybean			0.293 (0.085)	0.117 (0.115)			0.761 (0.302)	0.551 (0.433)
Cross-Price Coconut			−0.019 (0.054)				0.058 (0.086)	
Cross-Price Palm				−0.126 (0.095)				−0.066 (0.157)
Income	0.542 (0.175)	0.836 (0.412)	0.361 (0.347)	0.010 (0.424)	0.609 (0.179)	1.110 (0.253)	0.957 (0.926)	0.426 (0.573)

*Standard deviations in parentheses.

to changes in the prices of other edible vegetable oils. A higher own-price elasticity for soybean oil in the post-campaign period may be due to increased use of other domestic oils by food manufacturers, particularly those vegetable oils which have superior characteristics. Consumers led to perceive, as suggested by the ASA promotion, that soybean oil (at 15 percent saturated fat) provides a "better" bundle of characteristics than do tropical oils might then compare soybean oil unfavorably to other domestic or imported vegetable oils, such as canola or corn oil, in terms of saturated fat content.

The own-price elasticity of palm oil was 40 percent higher (in absolute value) in the post-campaign period, while the own-price elasticity of coconut oil declined moderately (Table 11.4). Cross-price elasticities of palm and coconut oils with respect to soybean and other domestic oils were, on average, much higher and more variable in the post-campaign period than before the tropical oils question in the media. The magnitude of the changes in cross-price elasticities were much greater for palm and coconut oils than for soybean and cottonseed oils, indicating that the ASA campaign generated greater repercussions on the imported tropical oils. Palm and coconut oils also became much more responsive to changes in income in the post-campaign period than before the promotions, reflecting their more limited uses in processes requiring specific taste or texture attributes.

Domestic soybean oil demand exhibited relative insensitivity to any changes in prices of tropical oils over the entire period (Figure 11.1). While the campaign served to educate consumers about the characteristics of various oils, it may have had an impact not foreseen by the ASA. Negative publicity about one or more competitive products can raise questions about relative health and safety factors for all products which are closely related. Embarrassingly, this may boost another competitor's products more than one's own. There is a suggestion in trends in the late 1980s that the ASA's campaign against tropical oils benefited corn and canola oils at the expense of soybean oil market share. Once educated about the fatty acid contents of various oils, consumers expressed increasing preferences for the oils lowest in saturated fats (i.e., canola and corn) over soybean oil.

The low estimated values of the persistence parameters suggest that the negative impact of the "fatty oils" campaign on palm and coconut oils may have been only a temporary structural change through 1988 (Figure 11.2). The effects of such generic promotions in an emotionally-charged consumer environment directly influence subsequent industry conduct, and, in the end, import–export relationships. Such domestic commodity promotion campaigns may have devastating short-term impacts on imports, which in turn are highly significant to the export markets of a small number of producing countries, in this case primarily Malaysia and the Philippines. Third-party reactions and the longer-term effects of actions of industry participants in an increasingly globally competitive environment may be more difficult to anticipate, but the food labeling legislation which resulted from the ASA campaign appeared to affect the U.S. consumption of tropical oils in a permanently negative manner.

References

Augusto, S., and P. Pollak. "Structure and Prospects of the World Fats and Oils Economy." *Papers Presented at the Workshop on Commodity Models and Policies (Vol. 1)*. Washington, D.C.: The World Bank, 1981.

Cooley, T. F., and E. C. Prescott. "Systematic (Non-Random) Variation Models Varying Parameter Regression: A Theory and Some Applications." *Annals of Economic and Social Measurement*. 2(1973a):463-73.

Cooley, T. F., and E. C. Prescott. "An Adaptive Regression Model." *International Economic Review*. 14(1973b):364-71.

Emery, W. L. "Soybean Oil and Its Competition." *CRB Commodity Year Book*. Jersey City, NJ: Commodity Research Bureau, 1985, pp. 21T-39T.

Fryar, E. and R. Hoskin. "U.S. Vegetable Oil Price Relationships." *Oilcrops Outlook and Situation*, U.S.D.A., December 1985 Yearbook.

Goddard, E. W. and A. K. Amuah. "The Demand for Canadian Fats and Oils: A Case Study of Advertising Effectiveness." *Amer. J. of Agric. Econ.* 71(1989):741-749.

Goddard, E. W. and S. Glance. "Demand for Fats and Oils in Canada, U.S. and Japan." *Canadian J. of Agric. Econ.* 37(1989):421-443.

Gudmunds, K. N., and A. Webb. *PS & D VIEW '91* [Computer file]. #89028. Washington, D.C.: Economic Research Service, U.S.D.A., December, 1991.

Judge, G. G., R. C. Hill, W. E. Griffiths, H. Lutkepohl, and C.T. Lee, *The Theory and Practice of Econometrics*, 2nd Edition, Wiley, New York, 1985.

Nyberg, A. J., "The Demand for Lauric Oils in the United States," *Amer. J. of Agric. Econ.* 52(1970):97-102.

Othman, J. B., C. S. McIntosh, and J. E. Houston. "Vegetable Oil Industry Conduct, A.S.A. Health Advertising and Impacts on U.S. Soybean Oil Demand." *The Changing U.S. Grain Marketing System: 1989 Proceedings of NCR-186, Performance of the U.S. Grain Marketing System*. Minneapolis, MN, October, 19-20, 1989, pp. 171-184.

Othman, J. B., J. E. Houston, and C. S. McIntosh. "Health Issue Commodity Promotion: Impacts on U.S. Vegetable Oil Demand." *Food Policy*, 18(1993)3:214-223.

Othman, J. B., C. S. McIntosh, and J. E. Houston. "Price Competition and Informational Product Differentiation Effectiveness: The Case of Edible Vegetable Oils in the U.S." FS #94-06, paper presented at the American Agricultural Economics Meetings, San Diego, 1994a.

Othman, J. B., J. E. Houston, and C. S. McIntosh. "Competition in U.S. Edible Vegetable Oil Markets: Persistence of Structural Changes." FS #94-08, paper presented at the XXII International Conference of Agricultural Economists, Harare, Zimbabwe, 1994b.

Reed, M. A., R. A. Ghaffar and A. Pagoulatos. "An Analysis of the Effects of Palm Oil Imports on the U.S. Soybean and Soybean Oil Industry." *Malaysian J. of Agric. Econ.* 2(1985):77-88.

American Soybean Association, *Soybean Digest,* September, 1987; November, 1987; December, 1987; and March, 1989, issues.

Soyatech Inc. *Soya BlueBook.* Bar Harbor, ME, 1986.

U.S. Department of Agriculture, *Agricultural Outlook,* Economic Research Service, USDA, Washington, D.C., May, 1988; May, 1989.

U.S. Department of Agriculture, *Food Marketing Review,* Economic Research Service, USDA, Washington, D.C., 1991.

U.S. Department of Agriculture, *Oil Crops Situation and Outlook Report.* Commodity Economics Division, Economic Research Service, Washington, D.C., various issues.

U.S. Department of Agriculture, *World Agriculture Situation and Outlook Report,* Economic Research Service, USDA, Washington, D.C., March 1987.

U.S. Department of Agriculture, *End Uses of Fats and Oils,* Unpublished Data.

U.S. Department of Commerce, International Trade Administration. *Foreign Economic Trends and Their Implications for the United States (Malaysia).* Washington, D.C., June, 1989.

U.S. Department of Commerce. *Survey of Current Business.* U.S. Dept. of Commerce, Washington, D.C., various issues.

Willimack, D.K. "U.S. Edible Fats and Oils Refining Capacities, 1983." *Oilcrops, Situations and Outlook Report.* U.S.D.A., Economic Research Service, May 1985.

12

U.S. Rice Milling Industry: Structure and Ownership Changes

Eric J. Wailes and Wayne M. Gauthier

Introduction

The U.S. rice milling industry has been in a constant state of adjustment since the mid-1970s. Changes in U.S. government rice programs, shifts in production patterns, export markets, and other factors have influenced changes in the size and location of rice milling in the United States (Cramer et al., 1990; Setia et al., 1994; Wailes and Holder, 1987). The objectives of this chapter are to: (1) describe the changing structure and ownership of the U.S. rice milling industry ; and (2) analyze the impacts of selected changes in the structure and ownership on the future performance of the rice milling industry.

The United States has been the second largest rice exporter in world rice trade over the past several decades. The majority of current rice milling capacity in the United States was built during a period when export markets absorbed approximately 60 percent of domestic output. Since 1980, however, U.S. per capita rice consumption has doubled. As a result, since the late 1980s, the domestic market has become more dominant relative to the export market. Over 55 percent of the milled rice from 1990 through 1995 was shipped to domestic markets.

The United States, with 2.6 million acres of rice, is a relatively small rice producer by world standards (less than 2 percent of world rice production). However, the U.S. rice industry has been one of the most dynamic over the past 25 years. U.S. rice planted area has increased by about 50 percent, and total production has doubled. While per capita rice consumption has declined in Asian nations such as Japan, South Korea, Taiwan, Thailand, and Malaysia, it has continued to increase in the United States.

The proliferation of high quality rice products created by the U.S. rice milling and food processing industries has been responsible for the increase in U.S. consumption. These products include many varieties of white milled rice, aromatics, pasta products, parboiled rice, snacks and candies, and flavored mixes. The United States has also become a significant importer of rice because of growing demand for specialty rice, especially aromatic types.

The United States currently exports about 45 percent of its production and has been, on average, the world's largest rice exporter after Thailand. Unlike all other exporters, the U.S. exports milled rice into all types of markets differentiated by quality and rice type (indica and japonica). U. S. rice exports have dominated and continue to command premiums over comparable Thai rice grades in the high quality export markets. These premiums reflect a preference for the consistent and timely delivery and quality of U.S. rice exports.

Number and Size of Rice Mills

U.S. rice milling capacity expanded between the mid-1970s and mid-1980s, following the elimination of acreage controls in 1974 and the rapid growth in rice exports beginning in 1973. The number of mills increased from 40 in 1972 to 44 in 1978 and to 66 by 1985 (Table 12.1). Most of the new mills were located in California, Texas, Mississippi and Arkansas. The number of mills in Louisiana actually declined, as several small mills closed.

An unanticipated decline in exports in the early 1980s resulted from a high U.S. domestic farm price floor (loan rate) and a significant dollar

Table 12.1 •   **Number of Rice Mills by State**

State	1972	1978	1985	1991	1994
Arkansas	8	8	21	16	13
Louisiana	17	16	12	8	8
Tennessee	0	0	1	0	1
Florida	0	0	2	1	1
Mississippi	1	3	6	2	2
Missouri	0	0	0	1	1
Texas	8	10	14	12	10
Southern Mills	34	37	56	40	36
California	6	7	10	9	7
Total U.S. Mills	40	44	66	49	43

Source: Wailes (1987) and Holder (1991); Childs; and Rice Journal(1995).

appreciation. Reduced exports resulted in reduced milled rice demand and higher per unit milling costs because of reduced mill capacity utilization. The higher milling costs were largely absorbed by millers because of the highly competitive nature of the milling industry.

Although U.S. rice exports increased again in the late 1980s and early 1990s, due to the marketing loan provision of the 1985 farm bill, a number of mills failed to survive this turbulent period. Between 1985 and 1991, 17 mills exited the industry, and an additional 6 mills exited between 1991 and 1994. Characteristics of mills that ceased operations included: (1) small to medium-sized mills, (2) locations in Louisiana and Texas, and (3) relatively export–dependent mills. In general, these firms were unable to successfully shift their product flow into domestic markets due to higher start-up costs, greater packaging expenses, and higher milling quality requirements. These firms were also characterized as having older facilities than the industry average. The cost of modernizing milling equipment and buildings to become competitive in the higher-quality name-brand domestic market was prohibitive. The exodus of these 23 rice mills caused an upheaval in the milling industry that resulted in reduced processing and distribution shares for the export-dependent milling regions. Most of Louisiana's remaining mills serve specific and limited niche markets such as aromatic varieties and brown rice. Even large firms in Texas that did not establish a greater domestic presence closed operating facilities in the 1980s.

By the early 1990s, milling had shifted away from the Gulf Coast (Louisiana and Texas) to the Mississippi Delta region and California. The share of milled rice shipments from regions less dependent on export markets, i.e., Arkansas–Missouri and California, expanded during the 1980s (Table 12.2). California mills had a greater market orientation than southern mills as they were less dependent on food aid (P.L.-480) shipments for exports. They were already well established in the domestic market, and provided milled rice to domestic food processors.

Farmer-owned cooperatives increased their share of total rice millings and shipments through the 1986–87 marketing year. But with further restructuring in the milling industry and the termination of ARI, a Texas cooperative, in 1988, the cooperatives' share of total shipments decreased. Cooperatives accounted for a larger share of the export market than of the domestic market prior to 1988–89. But since 1988–89, milled rice shipments from cooperatives have had a larger share of the domestic market than the export market.

Recent mill closures have eased the excess milling capacity situation in the 1990s. In early 1994, there were 43 mills which accounted for virtually all of the rice milled in the United States. The mills are located

Table 12.2 • Milling and Distribution Shares to U.S. Consumers of Direct Food Rice, Selected Years

Mill Area	1974–75	1984–85	1988–89
Arkansas	30.5[a]	40.4[a]	39.0[c,d]
Louisiana	10.1	16.1[b]	11.7[b]
Texas	47.2	30.2	25.3[c]
California	12.2	13.3	24.0

[a] Includes Mississippi mills.
[b] Includes Florida mills.
[c] Excludes Mississippi mills for Arkansas, included for Texas.
[d] Includes Missouri mills.
Source: Smith, Wailes, and Cramer (1990); Childs (1991).

in the rice production areas depicted in Figure 12.1. The mills are located by states as follows: 13 in Arkansas, 7 in California, 2 in Mississippi, 1 in Missouri, 8 in Louisiana, 10 in Texas, 1 in Tennessee, and 1 in Florida. The 43 mills are currently operated by 30 companies, of which 21 companies operate a single mill, 6 companies operate 2 mills, 2 companies operate 3 mills, and 1 company operates 4 mills. The mills that are part of a multi-mill operation are primarily located in Arkansas and secondarily in Texas. Milling in Louisiana and California is primarily performed by single-mill companies.

The U.S. rice milling industry is a heavily concentrated industry compared to other grains. The domestic market is currently more highly concentrated than the export market. The largest four rice milling companies have accounted for over 50 percent of domestic rice shipments since 1975–76. The top five milling companies have accounted for about 60 percent of domestic shipments since 1988–89. In terms of total (domestic and export) shipments, the top four (as well as top five) rice milling companies have accounted for a declining share of shipments since 1986–87 (Childs, 1992). This declining share is due to a drop in export shipments from the larger mills and an increase in domestic shipments from medium-sized mills with established national and niche markets. The closure or restructuring of two of the largest rice milling cooperatives has also reduced concentration ratios.

Product Value and Value Added

The rice milling industry converts rough rice into: (1) brown rice and hulls, or (2) white rice, bran, and hulls. A certain percentage of the product is further processed through parboiling, instantizing, or pre-cooking. The milling yield is typically 70 percent white rice, 8 percent bran, and 12 percent hulls.

Rough rice is typically milled once the moisture content has been reduced to 12 or 13 percent. Mills purchase either dried rough rice or green rice to be dried in their own facilities. The purpose of milling is to remove the hulls and bran from the dried rough rice and to produce a milled and polished (white) rice. The term "milling" does not have a concise and consistent meaning in the rice milling industry. Milling can either refer to the overall operations of a rice mill, i.e., cleaning, shelling, bran removal, size separation, etc., or it can simply refer to the removal of bran from brown rice to produce a whole grain white rice product.

Figure 12.1 • 1994 Location of U.S. Rice Mills

Milling by-products such as hulls, bran, and polish are directed to markets which make use of their unique characteristics. Food processors purchase brokens (less than two-thirds of a kernel) for use in food products such as beer, cereal, soups, and pet foods. Rice meant for direct consumption (typically with only a maximum of 4 percent brokens) is placed in individual consumer-size bags which range in weight from 1 to 100 pounds.

Until recently, rice hulls were typically treated as a waste by-product and disposed of in landfills. Today, rice mills utilize the hulls for a variety of end-use products including: livestock feed (60 percent), poultry bedding (15 percent), fuel for hull-fired boilers or process steam/heat for power generation (15 percent), and other various uses (10 percent) (Vellupillai et al., 1996).

Table 12.3 presents data on the volume, farm and mill value, and the value added by the U.S. rice milling industry. Volume milled has more than doubled over the past 25 years. A linear trend regression on volume milled indicates that, on average, an annual increase of 4 million cwt has been milled since 1970 (Table 12.4). Except for weak mill demand due to depressed export demand in the early 1980s (1982–85), mill output has increased. Weather and supply controls (such as the acreage reduction program, 50/92, and triple base (flex) mechanisms, implemented under the 1985 and 1990 U.S. farm bills) explain most of the year-to-year variation in production over the past ten years. Expansion in domestic and export markets since 1992 has resulted in higher prices and higher total farm and mill values.

Milling Costs

Wailes and Holder (1987) estimated per hundredweight (cwt) milling costs for various capacities of single- and multi-floor mills (Table 12.5). They assumed that mills with less than 500 cwt per hour capacity were single-floor establishments and mills with 500 cwt capacity and above were multi-floor facilities. Average costs decreased rapidly as mill size increased from 162 cwt per hour to 486 cwt per hour for single-floor mills. Decreasing average costs for increasing mill sizes were reported for mills that had full package lines (a 38 percent cost decrease from smallest to largest) as well as for those with more bulk packaging, i.e., 70 percent of packaging in cwt bags (a 40 percent cost decrease). Economies of scale were of smaller magnitudes for the larger multi-floor mills. In moving from a 567 cwt per

Table 12.3 • Volume of Rice Milled, Farm and Mill Value, and Value Added, U.S., 1970–95.

Year	Rough Rice Milling (mil. cwt)	Farm Value (mil. $)	Mill value[a] (mil. $)	Value Added[b] (mil. $)	Value Added[c] (mil. $)
1970	83.1	429.63	605.38	175.76	486.86
1971	94.1	502.49	692.67	190.17	505.78
1972	92.3	621.18	948.77	327.59	842.15
1973	90.2	1,244.76	1,973.04	728.28	1,763.39
1974	113.2	1,267.84	1,811.21	543.37	1,186.39
1975	98.6	823.31	1,279.64	456.33	912.66
1976	112.1	786.94	1,233.11	446.17	843.42
1977	112.4	1,066.68	1,741.92	675.24	1,199.37
1978	129.1	1,053.46	1,736.15	682.69	1,126.56
1979	137.9	1,447.95	2,195.38	747.43	1,107.30
1980	155.6	1,991.68	2,835.16	843.48	1,101.15
1981	150.6	1,362.93	2,171.24	808.31	956.58
1982	131.8	1,042.54	1,680.16	637.62	710.83
1983	125.2	1,072.96	1,677.01	604.05	652.32
1984	122.6	985.70	1,577.35	591.64	609.94
1985	124.5	812.99	1,434.44	621.46	621.46
1986	161.9	607.13	1,364.05	756.93	741.36
1987	152.6	1,109.40	2,173.31	1,063.91	1,003.68
1988	168.3	1,149.49	1,890.17	740.68	673.35
1989	159.3	1,170.86	1,842.23	671.37	582.79
1990	162.6	1,086.17	1,823.62	737.45	604.47
1991	161.8	1,226.44	1,930.25	703.81	555.93
1992	173.7	1,023.09	1,762.92	739.83	567.35
1993	176.7	1,410.07	2,530.74	1,120.67	834.37
1994	199.2	1,350.58	2,055.46	704.89	509.51
1995	188.2	1,590.29	2,548.63	958.34	674.89

Source: USDA, Rice Situation and Outlook, various issues.

[a] Based on 70 percent milled rice Arkansas, long grain No. 2, 8 percent Southwest Louisiana bran price, and 22 percent hulls valued at $6/ton.

[b] Nominal dollars.

[c] Value added from milling is deflated by CPI Index 1985=100.

Table 12.4 • Rice Mill Volume, Value Added, and Milling Margin Trend Regression Results, 1970-95.

Variable	Time Trend Coefficient	Intercept	R-squared
Volume of rough rice milled	3.95 (0.32)*	–7,691.95	0.87
Value-added			
Nominal	20.54 (4.37)	–40,047.6	0.48
Real	–17.02 (7.11)	34,562.6	0.19
Farm-mill margin[a]	–0.035 (0.0025)	9.56	0.56

* Standard error of the coefficient estimate.

[a] Observations are 1978–90 (monthly), real margins at constant 1982 dollars.

hour mill to a 1134 cwt per hour mill, multi-floor plant average costs per cwt for a full package line only declined from \$1.75 to \$1.52 (13 percent less), and for a 70 percent cwt bag plant from \$1.58 to \$1.35 (15 percent less).

Wailes and Holder also compared per cwt milling costs among five milling centers (Table 12.6). They estimated that Arkansas and California, the milling centers with the largest mill capacity, were the lowest cost millers of long grain with costs of \$2.51 per milled cwt. Higher milling costs were reported for milling centers in Mississippi at \$2.52, Texas at \$2.63, and Louisiana at \$3.05. Because California mills account for the bulk of medium grain milling (mills can generally process 20 percent more medium grain rice than long grain for the same mill capacity), California is the lowest-cost miller at \$2.30 per cwt, followed by Arkansas at \$2.40, Mississippi at \$2.41, Texas at \$2.52, and Louisiana at \$2.89. The descending order of milling cost competitiveness by milling region is California, Arkansas, Mississippi, Texas, and Louisiana mills.

Table 12.5 • Average Cost of Rice Milling, 1986

Mill size (cwt/hr)	Number of mills	Total mill capacity[1] (cwt/hr)	Percentage of total capacity	Full package line[2]		70 percent cwt bag	
				Avg. cost by mill size	Weighted[3] cost by capacity	Avg. cost by mill size	Weighted[3] cost by capacity
				$/cwt			
Single floor							
162	9	1,458	5.3	2.43	0.13	2.24	0.12
243	5	1,215	4.4	1.96	0.09	1.78	0.08
324	4	1,296	4.7	1.72	0.08	1.54	0.07
405	6	2,430	8.8	1.58	0.14	1.41	0.12
486	7	3,402	12.4	1.51	0.19	1.34	0.17
Multi-floor							
567	3	1,701	6.2	1.75	0.11	1.58	0.10
648	2	1,296	4.7	1.70	0.08	1.53	0.07
729	2	1,458	5.3	1.65	0.09	1.48	0.08
810	3	2,430	8.8	1.60	0.14	1.43	0.13
891	0	0	0.0	1.60	0.00	1.43	0.00
972	1	972	3.5	1.59	0.06	1.42	0.05
1053	4	4,212	15.3	1.55	0.24	1.38	0.21
1134	5	5,670	20.6	1.52	0.31	1.35	0.28
U.S. Total	51	27,540	100.0		1.66		1.48

Source: Wailes and Holder, (1987).

[1] Capacity is for white rice mills.

[2] A full package line distribution for these cost estimates is given in Wailes and Holder, (1987), Appendix table 30.

[3] Determined by multiplying each product line's average cost by its percentage share of total capacity.

Table 12.6 • Cost Comparison Among Six Mill Centers[1]

State	Mill capacity	Rough equivalent average milling cost	Milled equivalent cost			
			Long	Medium	Short	Weighted[2]
	cwt/hr	$/cwt	$/cwt			
Arkansas	7,695	1.50	2.51	2.40	2.32	2.49
California	6,966	1.29	2.51	2.30	2.57	2.37
Louisiana	5,184	1.81	3.05	2.89	2.80	2.99
Mississipi	1,701	1.51	2.52	2.41	2.33	2.52
Texas	5,670	1.58	2.63	2.52	2.44	2.63
U.S. Total	27,216	1.52	2.64	2.50	2.50	2.58

Source: Adapted from Wailes and Holder (1987).

[1] Assumptions: Mills run 245 days/year and two shifts/day. Output of mills is 70 percent packaged in cwt bags and 30 percent shipped bulk; milled equivalent is based on 96 percent head rice/cwt of milled rice. Capacity is for white-rice mills only.

[2] Milled equivalent costs are weighted according to the proportions of long-, medium-, and short grain rice milled in each state for the 1995 crop year.

Milling Margins

Rice mill marketing margins were calculated by subtracting the price received for rough rice from the f.o.b. milled rough equivalent price for Arkansas, Houston, and California for the period 1970 to 1992 (Figures 12.2–12.4). Estimates of the Arkansas and California margins are based upon their state average rough rice and mill prices. For the Houston estimates, the Texas average rough rice prices are used. The Houston and Arkansas mill price quotes are for long grain U.S. No. 2, 4 percent broken, and the California mill quote is for medium grain U.S. No. 1.

Arkansas and Houston milling margins move together and have nearly identical averages at $6.02 and $5.92 per cwt of rough rice, respectively. California margins are different from the southern states' milling margins because of the differences in rice type, industry structure, and location. Over the period reported, California milling margins have averaged $7.61 per cwt of rough rice. The higher California milling margins may

Figure 12.2 • Marketing Margins of Arkansas Rice Mills

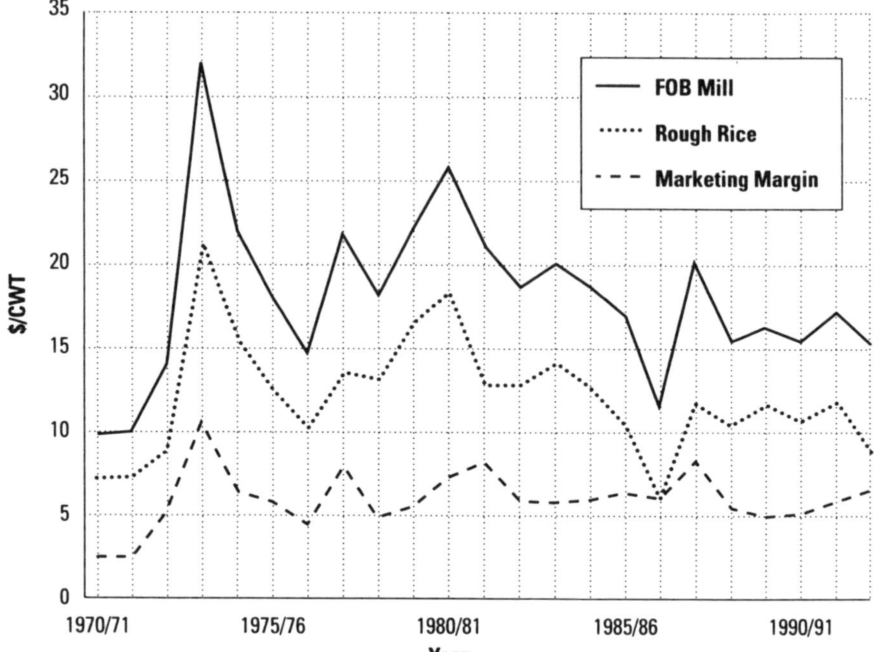

Source: USDA, Agricultural Marketing Service and National Agricultural Statistical Service

be attributed to higher costs due to the use of unionized labor. The higher returns to milling are due to relative market isolation and a lower degree of competition, the latter reflected by relatively fewer mills compared to the southern region.

In the early 1970s, the milling margins were fairly low, like mill prices, but then increased as the export boom began in 1972–73. Figures 12.2–12.4 indicate that milling margins are sensitive to changes in rough rice and f.o.b. mill prices—with millers able to increase their margins as prices increase and similarly decrease margins as prices fall.

A study by Chavez (1994) evaluated pricing efficiency in the U.S. rice market using a marketing margin approach. The average U.S. farm-mill real price margin (milled equivalent) for the period studied (1978–90) is depicted in Figure 12.5. Simple regression results for the U.S. farm-mill real price margin against a time trend generates a strong negative relationship, a decline in real milling margins by –3.5 cents per month or 42 cents per year (Table 12.4). This trend suggests that pricing efficiency or productivity has improved dramatically over the 1978 to 1990 period. Chavez evaluated three alternative farm-mill price margin models—markup pricing (MP), relative pricing (RP), and marketing cost (MC), to examine changes in pricing efficiency (Wohlgenant and Mullen, 1987; Gardner, 1975).

The MP model was found to best explain the behavior of price margins for U.S. rice.[1] The specification of the MP model is:

$$MARGINX1 = b_0 + b_1 WSALEPM1 + b_2 MCINDEX1 + b_3 BYPRODP1 + b_4 MONTHNO + b_5 Q_1 + b_6 Q_2 + b_7 Q_3 + e_1$$

where b_0 through b_7 are coefficients to be estimated, and

Figure 12.3 • Marketing Margins of Houston Rice Mills

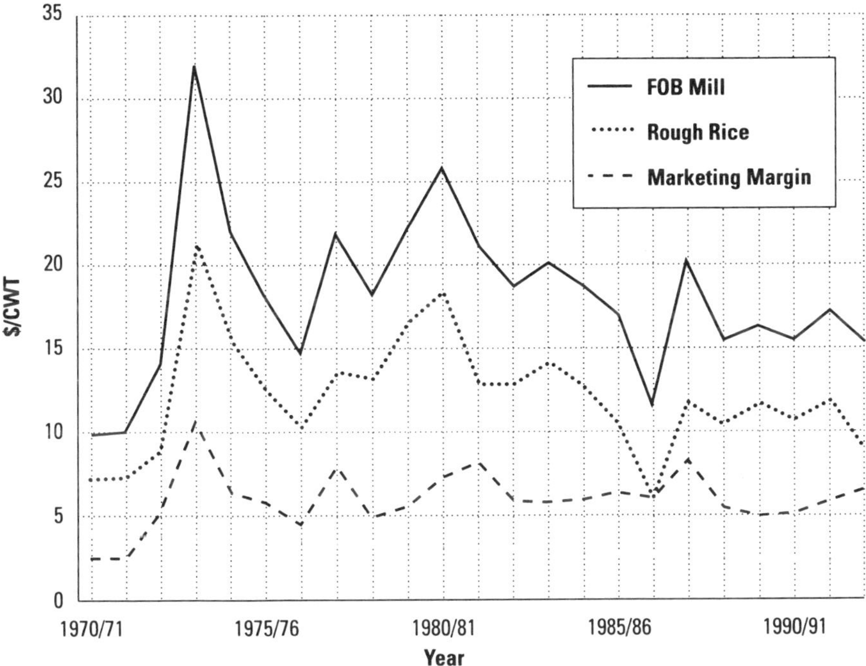

Source: USDA, Agricultural Marketing Service and National Agricultural Statistical Service

---

[1] The price margin models were evaluated using a combination of non-nested hypothesis test techniques, i.e. J-test (Wohlgenant and Mullen, 1987) and JA-test (Doran, 1993).

Figure 12.4 • Marketing Margins of California Rice Mills

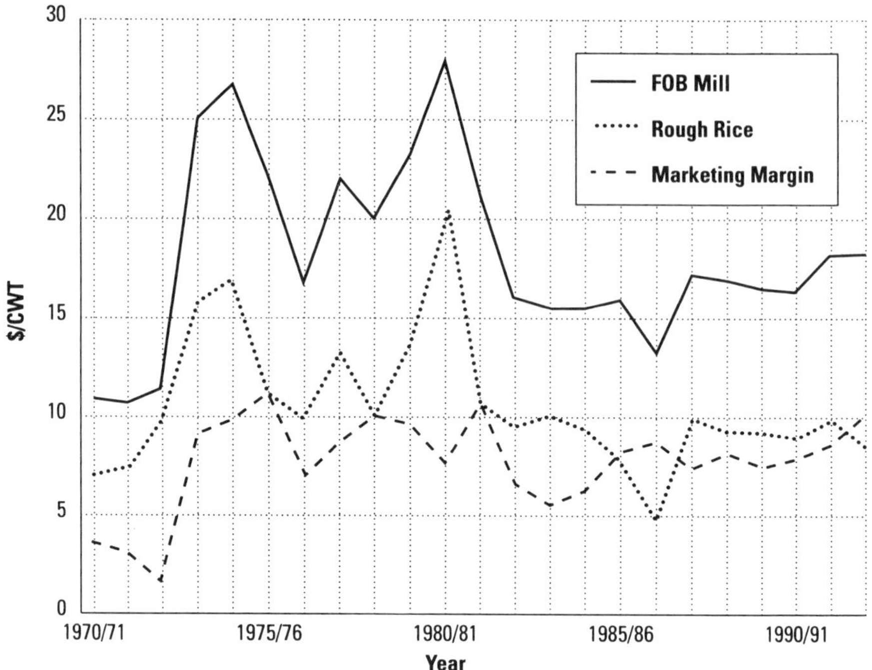

Source: USDA, Agricultural Marketing Service and National Agricultural Statistical Service

MARGINX1 = farm-mill price margin (f.o.b. mill price minus milled equivalent of farm price, in $/cwt).

WSALEPM1 = monthly average price for U.S. No. 2, milled rice, f.o.b. mill. Houston, Texas ($/cwt).

MCINDEX1 = index of marketing costs (simple average of hourly earnings of employees in grain mills, and producer price index of fuels and related products, and power).

BYPRODP1 = monthly average price of rice bran, f.o.b. mill, Southwest Louisiana ($/cwt).

MONTHNO = linear time trend (monthly).

Q_i's = quarterly seasonal variables (1 = Aug/Oct, 2 = Nov/Jan, 3 = Feb/Apr).

To analyze changes in efficiency of the U.S. rice market over time, the data were divided into three sub-periods: (1) 1978–1981, (2) 1982–1985, and (3) 1986–1990. These three sub-periods are hereinafter referred to as "Period 1," "Period 2," and "Period 3." These periods were selected to reflect significant differences in government rice programs. Period 1 reflects a significant expansion period following the elimination of the acreage allotments program in 1974. Period 2 reflects the time when government price floors priced the U.S. rice market out of world trade and drastic supply control mechanisms were used to limit the build-up of government-owned rice stocks. Period 3 is characterized by the implementation of the marketing loan program which decoupled rice demand from price supports and resulted in a rapid recovery in rice production, milling, and exports. The three-period analysis allows for observation of the marketing margin of U.S. rice through time as reflected by the variation in the relative fit and stability of the coefficients of the model across

Figure 12.5 • U.S. Rice Farm-Mill Price Margins, (milled basis, 1982$), 1978-90.

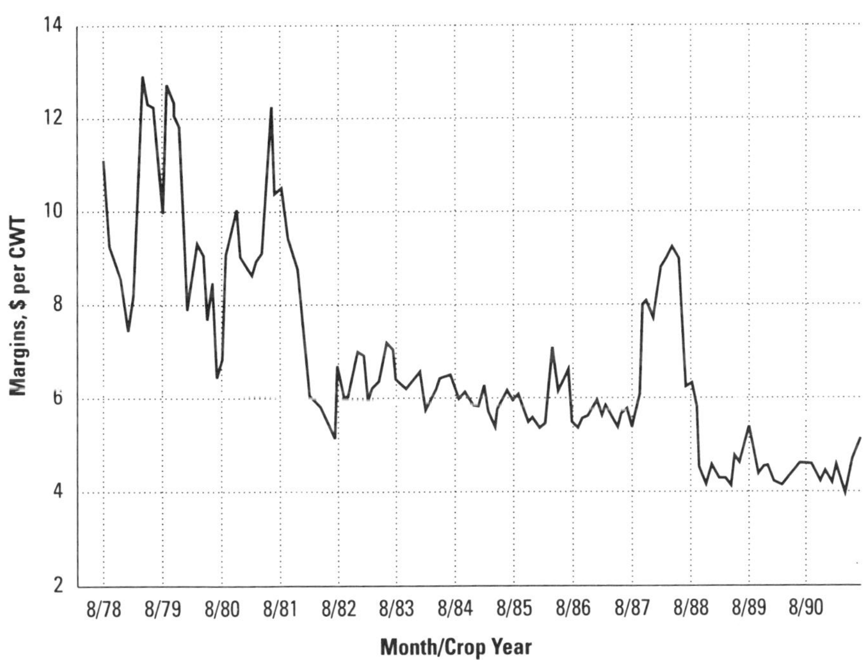

individual periods. The analysis serves as an indirect test of the hypothesis of no changes in marketing efficiency in the U.S. rice market.

The stability tests of the MP model across the three sub-periods rejected the null hypothesis at the 0.01 level of statistical significance. This result supports the validity of analysis for each sub-period and supports the idea that pricing efficiency in the U.S. rice market has changed over time.

A summary of the estimation results of the MP model by period is presented in Table 12.7. The relative changes in the pricing efficiency in the U.S. rice market were evaluated by analyzing changes in the relative explanatory power of the MP specification across the three periods. The MP model's explanatory power was highest during Period 3 as indicated by the model's distinctly superior statistics, i.e., highest R^2 and statistical significance of estimated coefficients (Table 12.7). The model explained 92 percent of the variation in the price margin during the period. For Period 1, the model managed to account only for 82 percent of the variation in margin. Period 2 showed the lowest R-square (0.37). One important feature of Period 3 statistics is that all of the coefficients of the major variables (mill price, marketing cost, and by-product price), including the linear time trend and one seasonal variable, are highly significant and have signs consistent with economic theory. These results contrast sharply with the Period 2 regression results which reflect poor explanatory and statistical significance for the primary economic variables that are hypothesized to determine the milling margins. This finding strongly indicates that the market forces operated more freely in the U.S. rice market during Period 3. One of the most obvious indications of improved efficiency is reflected in the decline in the real margins (means of the dependent variable in Table 12.7) for the three periods, i.e., $9.07 per cwt in Period 1, $6.19 in Period 2, and $5.46 in Period 3.

While this is not a direct test, these results provide evidence that pricing efficiency in the U.S. rice market has indeed improved significantly over the period of study. The level of efficiency dramatically improved during Period 3 compared to Period 2 based on the magnitude of the F-value and statistical significance of the coefficients. This indicates that the rice industry during Period 3 tended to be more market-oriented and efficient. This finding is consistent with theoretical expectations because of documented changes in the structure and the changing policy environment of the milling industry. A key policy change was the adoption of the marketing loan program, a relatively more market-ori-

Table 12.7 • Regression Results of a U.S. Rice Farm-Mill Markup Pricing (MP) Model

VARIABLE	PERIOD 1 (1978–81) n=48	PERIOD 2 (1982–85) n=48	PERIOD 3 (1986–90) n=60
CONSTANT	7.70	11.25	-2.36
	4.22[a,d]	6.32[a]	-0.77
WSALEPM1	0.50	-0.004	0.50
	10.49[a]	-0.04	26.13[a]
MCINDEX1	-0.17	-0.035	0.09
	-4.81[a]	-1.374[c]	2.19[b]
BYPRODP1	-0.43	-0.060	-0.63
	-2.17[b]	-0.506	-6.06[a]
MONTHNO	0.15	-0.017	-0.05
	4.17[a]	-2.47[a]	-4.09[a]
Q1	-0.04	-0.43	0.07
	-0.12	-2.82[a]	0.43
Q2	0.06	-0.44	0.16
	0.12	-2.36[a]	1.26
Q3	-0.24	-0.55	-0.35
	-0.65	-3.54[a]	-2.18[b]
Adjusted R2	0.82	0.37	0.92
F-value	698.55	1,751.16	1,430.64
Std. Error	0.86	0.36	0.41
S.S. Errors	29.45	5.26	8.66
Mean (Dep. Var.)	9.07	6.19	5.46
C. V.	0.09	0.06	0.07

Source: Chavez (1994).

[a] Significant at 0.01 level.
[b] Significant at 0.05 level.
[c] Significant at 0.10 level.
[d] t-statistics are below coefficient estimates.

ented program compared to the previous programs, which was implemented during Period 3 of this analysis.

Conclusions

The U.S. rice milling industry has operated in a very dynamic environment over the past two decades. Changes in U.S. government rice policies, instability in world rice trade, changes in technology, and milling costs have resulted in an uneven development in the rice milling industry over the past 25 years. This paper has identified the period of expansion during the 1970s, followed by a period of downsizing in the 1980s, and a current period entailing ownership restructuring.

The challenge to adjust to the market environment has become even greater as U.S. rice program policy is changed significantly under the Federal Agriculture Improvement and Reform Act of 1996, and specifically under Title I of this legislation, the Agricultural Market Transition Act. This change in policy will completely decouple income supports from production decisions and provide farmers with full flexibility in their planting decisions. The elimination of relatively high price and income supports, which have been tied to rice production in the past, are expected to result in a 10 to 20 percent reduction in the U.S. rice industry, with a larger impact expected in Texas and smaller reductions in California (Wailes et al., 1996). This magnitude of reduction will undoubtedly place considerable pressure on the U.S. rice milling industry, forcing continued downsizing and restructuring in favor of the more efficient milling firms.

The U.S. rice industry has improved pricing efficiency over time by making significant structural adjustments in the size, location, and operations of the milling infrastructure. As the rice industry moves into an environment of even greater market orientation and less government intervention, the U.S. rice industry is expected to experience further elimination of existing mills and a trend toward greater market concentration as fewer firms survive the challenges ahead. These findings are consistent with conventional economic theory which holds that competitiveness leads to efficiency.

The industry must improve efficiency if it is to maintain its share in the domestic and international markets. A continuation of the recent improvements in efficiency and competitiveness suggests that the U.S.

rice industry should be able to maintain a strong presence in the international rice market.

References

Chavez, Eddie C. 1994. *An Analysis of Pricing Efficiency and Competitiveness in the U.S. Rice Market*. M.S. Thesis. University of Arkansas.

Childs, Nathan. 1991. *U.S. Rice Distribution Patterns, 1988/89*. Econ. Res. Service, USDA, Statistical. Bulletin. Wash, DC.

Childs, Nathan. 1992. "Changes in the Structure of the U.S. Rice Milling Industry." mimeo. USDA, Economic Research Service.

Cramer, Gail L., Eric J. Wailes, Bruce Gardner, and William Lin. 1990. "Regulation of the U.S. Rice Industry, 1965-89." *American Journal of Agricultural Economics*, (November 1990): 1056-1065.

Doran, Howard. 1993. "Testing Nonnested Models." *American Journal of Agricultural Economics*, 75(1): 95-103.

Gardner, Bruce L. 1975. "The Farm-Retail Price Spread in a Competitive Food Industry." *American Journal of Agricultural Economics*, 53(3): 399-409.

Setia, Parveen, Nathan Childs, Eric Wailes, and Janet Livezey. 1994. *The U.S. Rice Industry*. USDA, ERS, Agricultural Economic Report No. 700.

Smith, Randell K., Eric J. Wailes, and Gail L. Cramer. 1990. *The Market Structure of the U.S. Rice Industry*. Arkansas Agricultural Experiment Station Bull. No. 921. Univ. of Arkansas Division of Agriculture, Fayetteville, Arkansas.

Vellupillai, Lakshman, Dean Mahin, Jamie Warshaw and Eric Wailes. 1996. *A Study of the Rice Husk-to-Energy Systems and Equipment*. Draft Final Report, Contract No. ACK-5-14228, NREL, U.S. Dept. of Energy, Golden, Colorado.

Wailes, Eric J., Gail L. Cramer, Eddie C. Chavez. 1996. *Arkansas Global Rice Model, International Baseline Projections for 1996*. Ark. Agr. Expt. Sta. Spec. Rpt. (forthcoming). Univ. of Ark. Div. of Agr., Fayetteville, Ark.

Wailes, Eric J. and Shelby H. Holder. 1987. *Cost Models and an Analysis of Modern Rice Mills*. Ark. Agr. Expt. Sta. Bull. No. 907. Univ. of Ark. Div. of Agr., Fayetteville, Arkansas.

Wohlgenant, Michael K. and John D. Mullen. 1987. "Modeling the Farm-Retail Price Spread for Beef." *Western Journal of Agricultural Economics*, 12-125.

Rice Journal. 1995. "1995 Rice Industry Guide." Raleigh, NC.

U.S. Dept of Agriculture. *Rice Situation and Outlook*. Various Issues. Wash. DC.

13

North American Malting Industry

William W. Wilson

Introduction

Barley is the third leading cereal crop grown in the world and the United States and the second most important crop (by volume) grown in the Canadian prairies. Barley production in the United States is concentrated in the northern plains states, contiguous to the principal growing regions in Canada. Both regions are surplus producers and, therefore, potential competitors. Traditionally, barley trade between these two countries has been negligible. However, recent changes in the policy, institutional, and competitive environments have resulted in increased trade and a rise in trade tensions. Institutional and policy factors appear to hold potential for further, drastic changes in competitive relationships and spatial flows.

This chapter describes the structural characteristics of the North American malting industry. The first section addressees issues on the supply of malting barley. The second section discusses malt demand. The structural characteristics of the malting industry are presented in the third section along with a discussion of strategic issues. The final section discusses some of the issues confronting U.S.–Canada competition in the malting barley sector.[1]

Malting Barley Production

Barley is classed into feed or malting varieties. In the United States, malting varieties are those that the American Malting Barley Association, Inc. (AMBA) has recommended for malting in specific states. Producers plant malting varieties in hopes of meeting malting quality standards (e.g., for plumpness and protein). If barley is of an approved variety, but fails to meet malting standards, it is sold as feed barley.

Breeding programs, agronomic practices, soil characteristics, and climatic conditions determine varietal types of barley grown in each state. Individual barley varieties were classified as feed or malting, 2-rowed or 6-rowed. The fraction of acreage planted to these individual barley varieties was multiplied by production in that state. Minnesota, North Dakota, and South Dakota grow primarily 6-rowed malting varieties. Since 1976, acres planted to 6-rowed feed varieties have been decreasing, while acres planted to 2-rowed feed varieties have been increasing in the Dakotas. Minnesota grows 6-rowed malting varieties almost exclusively.

Both feed and malting varieties are grown in Colorado and Idaho in significant amounts. The share of barley planted to 2-rowed malting varieties in Colorado has increased relative to feed varieties. In Idaho, the share of 6-rowed malting varieties has grown due to increased contracting with maltsters during the late 1980s. Montana and Wyoming mainly produce 2-rowed varieties. Most of the acres in Montana are planted to 2-rowed feed varieties, while 2-rowed malting varieties dominate in Wyoming. The coastal states of California, Oregon, and Washington produce mainly 6-rowed feed varieties. Planted acres in California and Oregon have been decreasing since 1962.

Malting varieties are grown predominantly in the Midwest states, with 6-rowed malting varieties accounting for approximately 55 percent of the acres grown and 2-rowed malting varieties accounting for approximately 12 percent in 1992. Area planted to 6-rowed malting varieties has increased more than have areas for 2-rowed malting and feed varieties.

Important institutional differences affect quality regulations in the United States and Canada. In the United States, the AMBA approves varieties for malting by state. Other varieties are classified as barley and are produced and sold for non-malting use. In the United States, both 6-rowed and 2-rowed malting varieties are white aleurone.

The variety approval process differs in Canada. Of particular importance is that varieties not approved for malting may still be produced, but only under contract (or for feeding purposes). Traditionally, all varieties approved for malting in Canada have been either 2-rowed white aleurone or 6-rowed blue aleurone. Other 6-rowed white aleurone varieties are grown under contract.[2]

The area planted to malting varieties indicates the proportion that is potentially available for malting. The proportion of the crop that is graded to be of malting quality, or sold for malting, is smaller. To put this in perspective, data were compiled on barley grade factors. Barley samples are collected annually in some states for quality evaluation. Some of those analyses estimate the proportion of crop graded as malting barley or as No. 3 or better malting barley.[3]

Minnesota and North Dakota malting barley typically grades No. 3 or better 60 percent of the time, and the South Dakota average is about 50 percent. Malting barley in Montana grades No. 3 or better about 30 percent of the time. Malting quality in Idaho is substantially greater than in the Midwest, reflecting the more controlled growing environment.

These figures were combined with other production data to derive estimates of malting barley production (graded No. 3 or better) for each of these states. Estimates were derived from the product of acres harvested, yield (including a lower yield for malting barley in Idaho), the percent of acres planted to malting varieties, and the percent of production graded No. 3 or better malting. As illustrated, North Dakota is the largest malting barley-producing state, with production of nearly 100 million bushels. Minnesota and Idaho each produce between 20 and 40 million bushels, and Montana and South Dakota each produce less than 10 million bushels.

Components of these figures are shown in Table 13.1, using averages for 1990–92. Estimates were not publicly available for Colorado, Montana, Oregon, Washington, and Wyoming. Figures used for these states were taken from discussions with industry participants. State production estimates were derived for malting barley, either graded No. 3 or better or sold as malting barley (depending on the reporting system).

For comparison, U.S. malt demand is about 104 and 40 million bushels for 6- and 2-rowed, respectively. Generally, the United States produces a large surplus of 6-rowed malting varieties graded No. 3 or better. In

Table 13.1 • U.S. Production Estimates Of Malting Barley

State	Harvest Acres (000)	Yield Malting bu/acre	6-rowed %	2-rowed %	Grade 3 %	State Production (000 bu) 6-rowed	2-rowed
CO	133.3	78	7	63	90	655	5,933
ID	763.3	74	26	21	91	13,247	10,988
MN	783.3	63	96	0	69	32,810	0
MT	1,410.0	46	4	34	80	2,076	17,382
ND	2,643.3	54	80	0	65	74,823	0
OR	151.6	68	4	2	80	346	132
SD	446.7	46	62	0	53	6,760	0
WA	466.7	56	2	9	90	447	2,140
WY	133.3	78	0	68	90	0	6,357
			average				
9-State Total	6,932	63	31	22	79	131,163	42,933

contrast, 2-rowed varieties are in relatively short supply. On average, production of 6-rowed malting barley is adequate, though periodic shortages of sufficient quality may occur due to weather vagaries. There is a greater chance of this occurring for 2-rowed malting barley than for 6-rowed malting barley. The relatively low level of 2-rowed malting barley suggests that this would be the type most likely imported.

U.S. Malt Demand and Utilization[4]

This section provides a description of the beer industry in the United States with particular focus on issues relating to malt consumption. Structural characteristics of the beer industry are presented, and malt utilization in the U.S. beer industry is analyzed.

Structural Characteristics of the U.S. Beer Industry

A few large firms dominate the U.S. beer industry. Anheuser-Busch is the largest with nearly twice the sales volume of the next largest firm, Miller. Table 13.2 provides a description of the structural characteristics of this industry in recent years. The number of beer companies has increased,[5] following many years of decline. This largely reflects the emergence of micro-breweries, a major trend in this industry. However, micro-breweries remain a small market segment (in comparison to the largest 20 breweries) in terms of sales and malt utilization.

Two important structural measures of this industry are shown in Table 13.2. First is the four-firm market share, a standard measure of concentration. This has increased in the past 10 years from 64 percent to 87 percent of sales. However, the Herfindahl Index is a more revealing measure of potential market power since it reflects the market share distribution. When there are few, similarly sized firms in an industry, competitive forces are greater than when there are few firms of disparate size; the Herfindahl Index captures this effect.[6] During the past 10 years, the Herfindahl index has increased from 1,617 to 2,818, reflecting an increase in the concentration of market power.

Malt Utilization

Beer consumption grew rapidly in the United States through the 1970s. During that period, beer production increased at about 4.6 percent annually. However, beer production has stabilized since 1980, while per capita consumption has decreased.[7] Reasons for the decrease include changing population demographics, drunken driving laws, and increasing health consciousness.[8]

Table 13.2 • Structural Characteristics of the U.S. Brewing Industry

	1981	1990	1991
Number of Beer Companies		142	169
Sales (mill barrels)		190	189
Average		1.34	1.12
Minimum (barrels)		166	77
Maximum (mill barrels)		86.5	86
Structural Characteristics			
Four-Firm Market Share %	64	86	87
Herfindahl Index	1,617	2,807	2,818

Source: Derived from data in *Modern Brewery Age*.

The growth rate of malt utilization in the U.S. beer industry has slowed, as has that for corn. Utilization of rice has increased. Estimated growth rates evaluated at 1991 for these three ingredients are malt, –.02 percent per year; corn, –.09 percent per year; and rice, +3 percent per year.[9] Thus, negligible growth in beer production and reduced utilization of malt have resulted in a declining domestic malt market.

The reduced utilization rates of both malt and corn in U.S. beer production and increased use of rice are attributable to many factors, including: (1) increased production and consumption of light beers, and (2) greater extraction rates from malt produced with new varieties. Malt comprised about 69 percent of ingredients in 1991. In Canada this figure is about 85 percent, implying utilization of 15 percent non-malt adjuncts.

Technical changes have encouraged ingredient substitution in beer production. These were estimated using a basic elasticity of substitution model, with trend term.[10] Estimated trend coefficients were –.03 and +.026 for malt/rice and malt/corn, respectively. Each coefficient was significant, and R^2's were .84 and .87, respectively. Trend coefficients measure changes in input utilization ratios through time. Malt utilization is decreasing relative to rice, but is increasing relative to corn.

Regional Estimates of Malt Demand

Beer production is geographically dispersed, but is generally concentrated near consumption regions. The geographical dispersion of production is shown in Figure 13.1. Some states are aggregated for disclosure reasons. The largest states (regions) for beer production and malt demand in 1991 were Arizona–Missouri–Idaho (reported as a single region), followed by Texas, California, and Georgia–Kentucky–West Virginia.

U.S. Malting Industry

Structural Characteristics

The malting industry is dominated by a few large privately held firms; however, their size differences are not as great as in the beer industry. Ladish has always been the largest maltster, followed in recent years by Anheuser-Busch, Fleischman-Kurth (a division of ADM), Great Western, and Froedtert. Others are small, typically operating single-plant units.[11]

Structural characteristics of the North American industry are shown in Table 13.3. Industry capacity increased rapidly during the 1970s, gen-

Table 13.1 • Beer Production by State

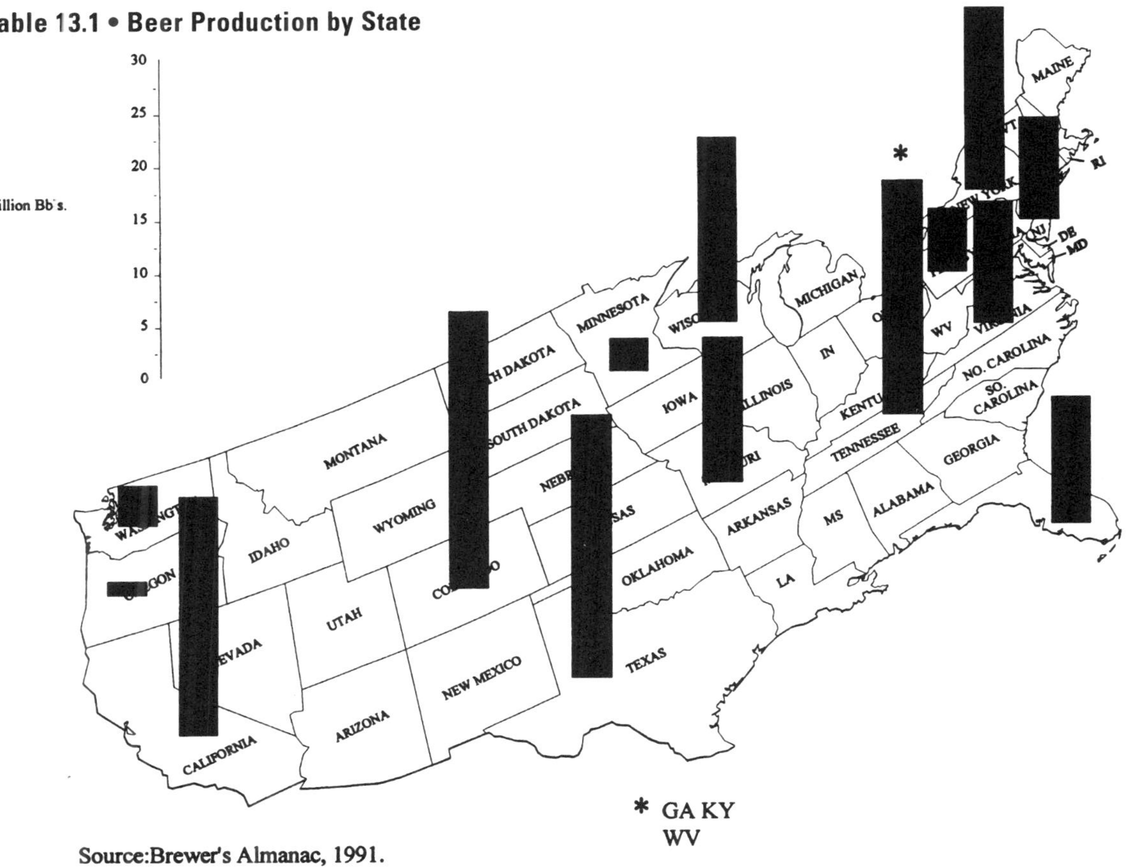

Source:Brewer's Almanac, 1991.

Table 13.3 • Structural Characteristics of the North American Malting Industry

Item	1968 U.S.	1980 U.S.	1992 U.S.	1992 Canada	1992 North America
Number of Plants	na	37	23	6	29
Number of Firms	31	26	13	4	16
Industry Capacity					
(million bu malt)	96	178	186	39	225
Structural Characteristics					
Brewer Controlled (%)	16	20	27	59	43
Four-Firm Market Share (%)	37	51	59	100	60
Herfindahl Index	687	1,009	1,208	4,010	1,178

erally coinciding with increases in malt demand by the beer industry. However, growth in capacity was minimal during the 1980s. The largest four firms control about 59 percent of capacity, up from about 37 percent in 1968. Brewer-controlled capacity has also increased from 16 percent in 1968 to about 27 percent in 1992.

The Herfindahl Index rose from 687 in 1968 to 1,208 in 1990, indicating increased concentration in the malting industry. However, the largest firms in the industry are of similar size, which intensifies competition in sales and procurement. For comparison, the Herfindahl Index calculated for Canada is 4,010. If the malt industry is viewed within a combined North American geographical market, structural characteristics are similar to those for the United States alone.

Malting Capacity by State/Region

The geographic distribution of malt plants is illustrated in Figure 13.2 with comparisons to malt demand by breweries. Historically, the industry developed in Wisconsin and Minnesota, and these states still dominate the industry. In the early 1970s, Ladish expanded with a new plant in North Dakota. Other new capacity in recent years has been added in Idaho to capture the expected growth in west coast demand for beer and barley production in those states.

Figure 13.2 • North American Malting Plants and Barley Production

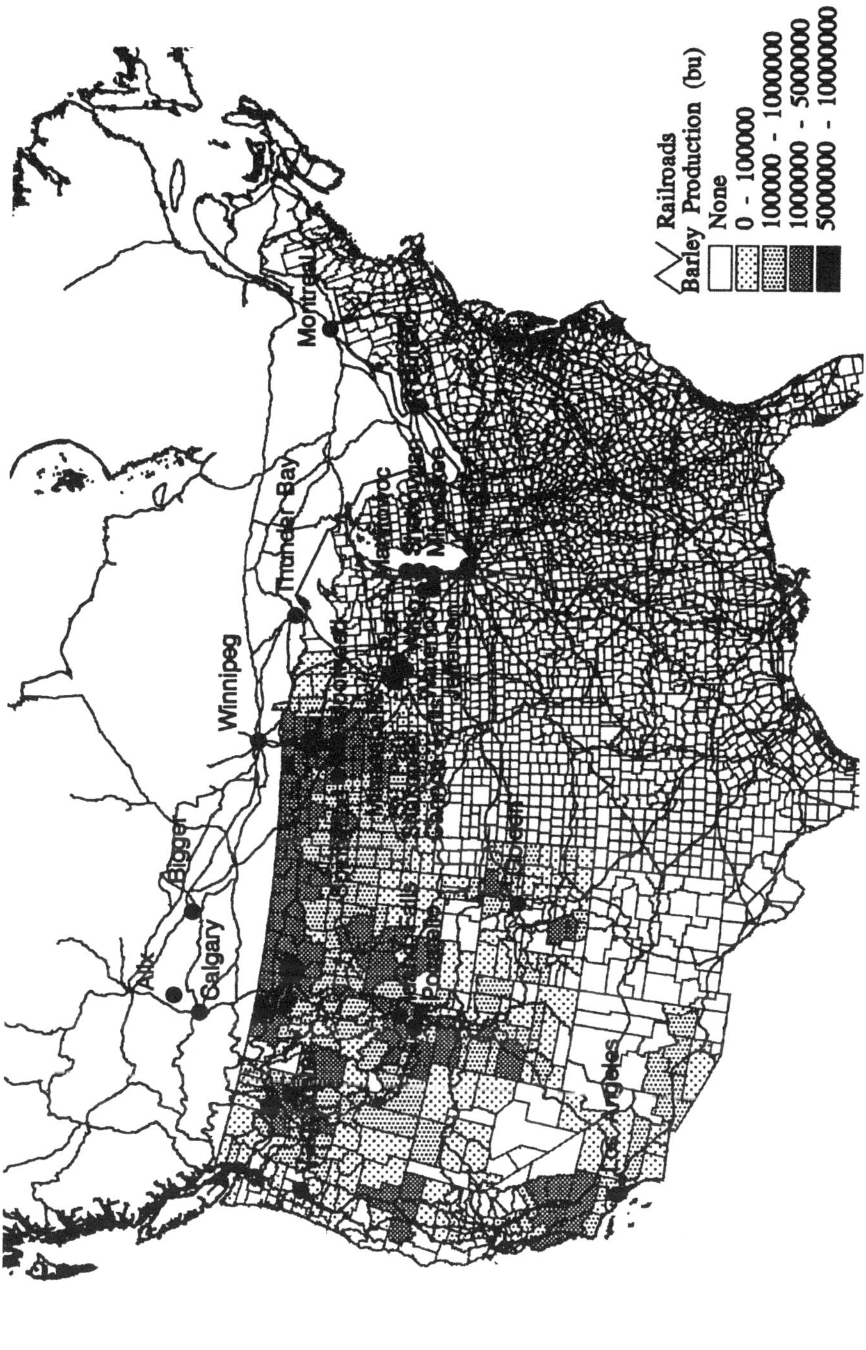

Firm Strategies

The U.S. malting industry is relatively concentrated. However, demand is stagnant, and excess capacity exists. At least four elements of broader firm strategies are apparent:

Globalization • The U.S. industry has been dominated by large single-plant private firms. However, over the past decade numerous changes generally have reflected a more global industry.

Cargill entered into a joint venture with Ladish in 1991. This represents Cargill's first entry into the U.S. malting industry, though it remains the largest maltster company in France and the world. Froedtert, previously owned by Harvest States Cooperative, was sold to a consortium of firms, who eventually sold to a large French malting firm.

At least three significant changes have occurred within North America in the past few years: (1) Great West Malting Company, previously a subsidiary of Univar, was acquired by Canada Malting, the largest malt company in Canada, making Canada Malting the largest malt firm in North America; (2) more recently Canada Malting was acquired by ConAgra; (3) Schrier purchased 51 percent of Prairie Malt (along with 42 percent by Saskatchewan Wheat Pool and 7 percent by employees) in September 1989; (4) ADM, which owns Fleischman Kurth, the second largest non-integrated maltster in the United States, bought 65 percent of Dominion Malt in September 1990[12]; and (5) Rahr has entered into a joint venture to build a new plant in Alix, Alberta.

Vertical Coordination • Before the early 1970s, brewery firms were not extensively integrated into malting in the United States. However, beginning with the expansion of Anheuser-Busch into malting in the early 1970s, inter-firm rivalry has intensified. Anheuser-Busch and Coors are extensively integrated into malting. Other brewers have malt plants (e.g., Miller, Genesea, Stroh), but on a lesser scale. In 1992, 28 percent of the capacity in the United States was owned and operated by brewery firms, up from about 16 percent in 1970. In addition, a common practice is for brewers to enter into "toll-malting" arrangements.[13]

Some brewers are also actively involved in the country elevator business, farmer contracting, and variety development. Anheuser-Busch began integrating into the country elevator business in 1983. All brewers are involved in variety development through their industry associations. However, Anheuser-Busch and Coors each have seed companies to develop varieties, and to clean and distribute seed to growers under contract.

Procurement Strategies • An important element of strategy is barley procurement. Two of the top three brewers have been actively involved in contracting programs. Anheuser-Busch has been involved in breeding programs since the early 1970s. Coors develops its own varieties and has contracted with producers for virtually all of its needs. However, Coors has recently announced cuts in its contracting program, and Anheuser-Busch has changed its approach to contracting.

Exports • Until recently, the United States has not been an active exporter of malt. However, the U.S. malting industry is pushing to expand exports with assistance from the Export Enhancement Program (EEP) (see below).

Industry Capacity Utilization Rates

Table 13.4 shows elements of supply and demand for capacity in the U.S. domestic malt industry[14] with comparisons to Canada.[15] Imports and exports are minimal, though exports have grown rapidly in the past several years. Exports comprised only about 5 percent of 1992 malt production in the United States, compared to 41 percent in Canada. Capacity utilization in the U.S. industry is about 81 percent, down sharply from the late 1960s.

Important points demonstrated in this table are: (1) there is a greater use per barrel in Canada; (2) exports comprise a greater share of production in Canada than in the United States; and (3) capacity utilization is somewhat greater in Canada than in the United States.

U.S. Malt Trade

The United States has routinely imported malt since the mid-1940s. In some years, imports were as great as 60,000 mt. However, malt imports have been declining since the early 1980s. Virtually all malt imports are from Canada. The United States has also been an exporter of malt since the 1940s. However, the volume decreased from about 100,000 mt per year in the late 1940s to less than 20,000 mt in the late 1970s. Since then, malt exports have expanded rapidly, nearly tripling during the 1980s.

The largest markets in recent years are Mexico, the United Kingdom, Japan, and the Philippines. Numerous other small markets exist, primarily in the Caribbean and South America.

Table 13.4 • Comparative Statistics: U.S. and Canadian Malting

		United States			Canada	N. Am
	Units	1968	1980	1992	1992	1991
Malt Beverage Production	mill brl	117	188	203	17	220
Use Per Barrel	lb/brl	28.2	26.7	24.0	39.5	
Domestic Malt Used	mill bu	97	148	144	19.5	164
Malt Export	000 mt	36	39	122	205	
Malt Import	000 mt	36	45	18		
% Production Exported		2	2	5	41	
Total Malt Production	mill bu malt	97	148	150	32.8	183
Total Capacity	mill bu malt	96.05	178	186	39	225
Capacity Utilization		1.00	0.83	0.81	0.84	

Sources: Derived from data contained in *Brewers Almanac* (1992); U.N. FAO "Commodity Trade Statistics," *Trade Yearbook* (various years), Canada Grains Council *Statistical Handbook*, (1992): and Canadian Wheat Board (December 1992). Industry capacity in Canada includes Westcan Malt in Alix, Alta.

China is a major importer and potential growth market for malt and malting barley. However, as is true for many malt importers, China has a strong preference for 2-rowed varieties.[16] U.S. maltsters have been working with some Chinese brewers. In general, these brewers are "impressed and comfortable" with U.S. 2-rowed barley, although they "still have some concerns about 6-rowed barley (primarily the fear of higher protein levels)" (*Barley Bulletin*, December 1992).

Malt has been targeted in numerous EEP initiatives, beginning in 1986. Since then, nearly 300,000 mt of malt have been sold through this program. The Targeted Export Assistance (TEA) program has also been used to promote U.S. malt exports. Sample shipments are sent to buyers, primarily in Latin America, to familiarize them with U.S. malt varieties.

A summary of the use of EEP in the U.S. barley sector is shown in Table 13.5. All figures are for the crop years from 1985–86 through 1991–92. Barley sales under EEP during this period were 11.4 mmt, compared to 257,000 mt of malt. Barley sales under EEP have a greater chance of being consummated relative to those of malt, with 78 percent for barley versus 33 percent for malt.[17] The average bonus on malt is nearly three times as great as that for barley. Barley sales ranged from 397,000 mt in 1985–86 to a high of 2.9 mmt both in 1986–87 and 1987–88. Malt exports increased from 43,919 mt in 1985 to 122,244 mt in 1991. From 1985–86 to 1991–92, percentages of U.S. exports sold under EEP have been 84 percent and 41 percent, respectively, for barley and malt (Table 13.5).

In October 1992, breaking convention, Colombia (50,000 mt) and Turkey (100,000 mt) were targeted specifically for "malting barley;" however, no sales were consummated. Targeted countries for barley include Algeria, Bulgaria, Cyprus, Hungary, Iraq, Israel, Jordan, Malta, Morocco, Poland, Romania, Saudi Arabia, Switzerland, Tunisia, Turkey, and the former Soviet Union. Targeted countries for malt include Algeria, Brazil, Burundi, Cameroon, Caribbean countries, Central American countries, Colombia, Iraq, Nigeria, Peru, Philippines, and Venezuela. Important non-EEP customers include Denmark, the United Kingdom, and Japan.

Pressures Affecting North American Barley Trade

One of the more important issues confronting the North American malting sector is pressure to evolve toward a set of policies more compatible with free trade. Pressures giving rise to the trade tensions and other differences between these systems are discussed briefly.

Table 13.5 • Summary of EEP in U.S. Barley Sector (1985–86 to 1991–92)

	Barley	Malt
Initiatives (000 mt)	14,650	780
Sales Under EEP (000 mt)	11,436	257
Sales of Initiatives (%)	78	33
Total Exports (1985–86 to 1991–92) (000 mt)	13,586	625
Sold Under EEP (%)	84	41
EEP Bonus Weighted Average ($/mt)	$32	$97

Derived from USDA data sources.

Trade and Agricultural Policies

The United States and Canada have different agricultural policy mechanisms, and different regulations governing grain trade. These have important consequences for North American barley.

The Conservation Reserve Program (CRP) and other supply control mechanisms have affected barley production in the United States. Up to 20–25 percent of U.S. barley base acres have been idled under CRP, while the Acreage Reduction Program (ARP) set-aside has averaged 11 percent of base acreage over the past decade. EEP has played a major role in assisting barley exports, and more recently exports of malt and malting barley. This program has the effect of raising domestic prices, creating price spreads (net of transportation) that favor sales of Canadian barley into U.S. markets. These price spreads are apparent at interior shipping points.[18]

Another important element of U.S. trade and agricultural policy is Section 22 of the Agricultural Adjustment Act. This provides a mechanism whereby the Secretary of Agriculture can impose quotas on imports of a commodity if they adversely affect domestic farm program operations. Technically, either an *ad valorem* import duty of 50 percent or import quotas (not to exceed 50 percent of a representative movement) could be imposed, either through an emergency action or following study by the U.S. International Trade Commission (National Grain Trade Council, 1993, p. 1).

Canada's policy mechanisms and marketing system differ drastically. Unlike the United States, Canada does not restrict the production of individual grains through explicit government intervention. Acreage has not been idled through government policies since the LIFT (Lower Inventory for Tomorrow) program of 1971.

The Canadian Wheat Board (CWB) is the fulcrum of many of the marketing and export policies in the Canadian system. As a single-seller agency,[19] the CWB can target markets, execute longer-term marketing strategies, and discriminate among customers using price and other marketing policies. This general approach to marketing is facilitated by having a monopoly on procurement and by being able to control imports and exports. However, the CWB competes with non-Board sales of barley and on-farm feed use.

Another important feature of the Canadian system is that subsidies are paid directly to railroads for a portion of the shipping cost. While this has a long and controversial history within Canada, rail subsidies have recently become a focal point of trade disputes with the United States.[20] To some extent, U.S. pressure may have contributed to Canada's movement toward reforming the "method of payment" of the Crow subsidy.

In June 1993, the Minister of Agriculture announced major reforms of Canada's marketing system. The proposed "Continental barley market" would have removed the CWB's monopoly over North American sales. Canada's export licensing scheme was to be eliminated, allowing Canadian producers unrestricted access to U.S. markets.[21] The reforms were implemented in August, but were soon reversed through court action. For now, the Canadian Wheat Board retains control over all barley exports.

The U.S.–Canada Free Trade Agreement (CUSTA) has no unique features for barley or malt trade. Tariffs are to be reduced over a 10-year period. Rail subsidies are not allowed on barley shipments to the West Coast, but are allowed for easterly movements. The United States retains Section 22 authority. Canada will retain import licensing authority for barley and malt until Producer Subsidy Equivalents (PSE) became equal in the two countries.

The North American Free Trade Agreement (NAFTA) contains a few special features regarding barley. Specifically, quotas for exporting to Mexico are allocated to Canada and the United States; sales within these quota limits are tariff-free.

Policies and marketing mechanisms differ substantially between the United States and Canada. Without possibilities for bilateral trade, this would pose no special difficulties. However, for different reasons, an open trading environment is seen as incompatible with existing institutions and policies on both sides of the border.

Market Dynamics

Barley markets have important underlying dynamics and spatial dimensions. Beer demand has grown slowly. In fact, U.S. beer consumption has fallen on a per capita basis, resulting in a growing surplus of malting capacity in the United States. However, the international market for malt has been growing.

The West Coast region is one of the largest feed deficit markets in North America. This market is traditionally served with midwest corn and barley, both of which incur relatively high freight costs. Suggestions have been made that this market could be competitively served from Canada, particularly Alberta, which is the primary barley production region in Canada.

Marketing Policies

Numerous marketing policies affect the barley sector. First, major differences exist in barley quality. U.S. and Canadian supplies are heterogeneous (and subject to much regional variation) in terms of varieties planted and crop quality. End-use requirements vary across brewers, which affects the demand for specific barley varieties and malt. Second, this industry is subject to a fairly high degree of vertical integration, as some brewers have acquired malt plants. Third, handling tariffs in the two countries reflect different competitive environments. Complicated rail and truck freight rate structures exist, resulting from competitive and structural attributes within each country.

Transportation and Handling Costs

There are numerous reasons foe what we refer to as prairie-border-crossing trade. These include spatial price differentials and the system of delivery quotas in Canada. Restrictive delivery quotas and the demand for cash can induce prairie-border-crossing shipments by Canadian shippers or producers. Spatial price differentials are affected by a number of factors, including quality premiums, EEP subsidies, and transport and handling costs. Because of fundamental differences in marketing institutions and philosophies. Canadian initial payments (and therefore "street prices") can differ drastically from U.S. spot prices.

One of the major differences in the marketing systems between these two countries concerns grain handing and transportation. There are important structural differences, as well as differences in regulatory environments, between the United States and Canada.

Most notable are the Western Grain Transportation Act (WGTA) rail rates and the related subsidy mechanism. This results in a pricing regime whereby the shipper pays a portion of the total rail shipment cost, referred to as the "shipper's portion." The balance, the "government's cost," is paid directly to the railroads as a subsidy. These levels and proportions are adjusted on an annual basis.[22]

These rates apply to all rail movements of barley and malt from the Prairies to Vancouver (for offshore export) and Thunder Bay. Under terms of the CUSTA, exports to the western United States do not qualify for subsidized rail rates; however, the rates do apply for shipments through Thunder Bay to the eastern United States. A proposal was made to change these rates, beginning in the 1994–95 crop year. The rates were increased in 1995. Specifically, railroads will be allowed to raise the shipper portion of the rates charged on WGTA movements, resulting in lower producer prices in the Prairies. However, rates were frozen to the year 2000 pending review.

Notable differences also exist in the handling systems in the two countries.[23] In Canada, the Canadian Grain Commission establishes handling regulations for licensed elevators and maximum tariffs for each function (e.g., storage, country handling, cleaning, fobbing). In contrast, a multitude of competitive forces in the United States determine rail rates and handling costs. Generally, these have resulted in lower handling costs.

Recent Studies

Opportunities for North America barley trade have inspired much debate in Canada. Recent studies have reached sharply different conclusions about whether the Board should retain its monopoly over Canadian exports. These studies provide numerous analyses of the North American market for barley and competitive pressures giving rise to change. They are not reviewed here but are listed in the references.

References

Agriculture Canada. September 1992. *Grains and Oilseeds: Regulatory Review* . Ottawa, Ontario.

Alberta Agriculture. *A Proposal for a North American Continental Market for Barley*. Edmonton, Alberta.

American Malting Barley Association, Inc. Issues from 1960 to 1992. *Know Your Barley Varieties*. Milwaukee, WI.

Ash, M., and L. Hoffman. December 1989. *Barley: Background for 1990 Farm Legislation*. Washington, DC: U.S. Department of Agriculture/Economic Research Service/Commodity Economics Division.

Beer Institute. 1992. *Brewers Almanac 1992: The Brewing Industry in the United States*. Washington, DC.

Brewers Association of Canada. *Annual Report*. Various Years. Ottawa, Ontario.

Brewing and Malting Barley Research Institute. *Barley Briefs*. Selected issues. Winnipeg, Manitoba.

Canada Grains Council. 1992. *Canadian Grains Industry Statistical Handbook*. Winnipeg, Manitoba.

Canadian Wheat Board. December 1992 (revised). *Performance of a Single Desk Marketing Organization in the North American Barley Market*. Winnipeg, Manitoba.

Carter, C. March 31, 1993. "An Economic Analysis of a Single North American Barley Market." Report submitted to the Associate Deputy Minister, Grains and Oilseeds Branch, Agriculture Canada. Davis, CA.: University of California-Davis.

Department of Cereal Chemistry and Technology. *Regional Crop Quality Report*. Various years. Fargo: North Dakota State University.

Food and Agricultural Organizations of the United Nations. *FAO Trade Yearbook*. Various years. Rome.

Interstate Commerce Commission. 1991. *Public Use Waybill Data*. Computer tape distributed by ALK Associates. Princeton, NJ.

Johnson, D., and B. Varghese. August, 1993. "Estimating Regional Demand for Feed Barley: A Linear Programming Approach." *Agr. Econ. Report No. 303*. Fargo: North Dakota State University, Department of Agricultural Economics.

Johnson, D. and W. Wilson. December 1994. "North American Barley Trade and Competition". *Agricultural Economics Report No.xx*, Department of Agricultural Economics, North Dakota State University, Fargo.

Magnusson, G., and M. Lerohl. 1991. "Alternative Alberta Barley Trade Scenarios With the Northwest United States." Project Report. University of Alberta, Department of Rural Economy.

National Grain Trade Council. October 27, 1993. *Wheat Imports: Should they Be Limited?* Washington, DC.

North Dakota Barley Council. *Barley Bulletin*. Selected issues. Fargo, ND.

Satyanarayana, V., W. W. Wilson, and D.D. Johnson. "Import Demand for Malt: A Time Series and Econometric Analysis," Dept. of Agricultural Economics, North Dakota State University, Fargo, ND., *forthcoming*.

Satyanarayana, V., W. W. Wilson, D. D. Johnson, and F. J. Dooley. "World Malting Barley and Malt: Competition, Marketing and Trade", Dept. of Agricultural Economics, North Dakota State University, Fargo, ND., *forthcoming*.

Schmitz, A., R. Gray, and A. Ulrich. April 1993. *A Continental Barley Market: Where are the Gains?* University of Saskatchewan, Agricultural Economics Department.

Takayama, T., and G. Judge. 1971. *Spatial and Temporal Price and Allocation Models.* London: North-Holland Publishing Co.

U.S. Department of Agriculture (F), National Agricultural Statistic Service/Agricultural Statistics Board. *Crop Production.* Various yearly summaries. Washington, DC.

U.S. Department of Agriculture (G), ERS. Various reports. *Feed Situation and Outlook Yearbook.* Washington, DC.

U.S. Department of Agriculture. "PS&D View," Commodity database. Washington, DC: Economic Research Service.

U.S. Department of Commerce, Bureau of the Census. June 1990. *1987 Census of Agriculture.* Volume II, Subject Series, Part 1.

Veeman, M. "A Comment on the Continental Barley Market Debate." *Canadian Journal of Agricultural Economics,* Vol. 41(3) 1993:238-287.

Weslowski, R. 1995. *Malt Import Market Characteristics,* MS Thesis submitted to the Dept. of Agricultural Economics, NDSU, Fargo, ND.

Wilson, W., D. Scherping, D. Cobia and D. Johnson. December 1993. *Economics of Dockage Removal in Barley: Background, Cleaning Costs, Handling, and Merchandising Practices.* Fargo: North Dakota State University, Department of Agricultural Economics.

Wilson, William W. June 12, 1993. "Impacts of Canadian Grain Imports on U.S. Markets." Testimony before the Subcommittee on General Farm Commodities, Committee on Agriculture, U.S. House of Representatives.

Wilson, William, W. May 1985. "Production and Marketing in United States and Canada." in *Barley,* D.C. Rossanussen, Ed., ACS 325.8, pp. 483-510, American Agronomy Association, 1985.

Wilson, W. W., and Johnson, 1995. "North American Malting Barley Trade: Impacts of Difference in Quality and Marketing Costs." *Canadian Journal of Agricultural Economics* (forthcoming).

---

[1] Greater detail on all of these topics is presented in Johnson and Wilson (1994).

[2] Carter (1993, p. 9-10) documents the percent of barley sold for malting (p. 10) and discusses factors affecting variety planting decisions.

[3] Non-grade factors (such as protein) may also determine a particular shipment's acceptability for malting.

[4] The Canadian beer and malting industry is discussed in Carter (1993, pp. 13-14).

[5] The data reported in Table 13.4 are firm data, not plant data. Many of the largest firms shown in Table 13.4 are multiplant brewers.

[6] The Herfindahl index is derived as

$$H = \sum S_i^2 * 10,000$$

where S_i is the market share of firm i. H ranges from 0, reflecting a perfectly competitive market, to 10,000, reflecting a pure monopoly.

[7] See Weslowski (1995), Satyanarayana and Wilson for a detailed analysis of growth rates in beer production for the United States, Canada, and 138 other countries.

[8] Similar trends have occurred in Canada. See the annual reports of the Brewers Association of Canada for a graphical depiction.

[9] Derived from regression of total utilization of each ingredient in beer production from 1964 to 1991, using a regression model of the following general form: $U = \alpha + \beta_1 T + \beta_2 T^2 + \beta_3 T^3$ where U is total utilization of each ingredient and T is a time trend with T = 1 in 1968.

[10] To evaluate substitutability among these ingredients in beer production, a basic elasticity of substitution model was estimated, using these prices and utilization levels. The model was specified as: $log(U_m/U_i) = a_i + \beta_i[log(P_m/P_i)] + \beta_t[logT]$ where U and P are utilization and price, respectively, m is for malt, i is for other inputs, and T is a time trend. In this basic model, β_i is the elasticity of substitution, measuring the response of input ratios to changes in relative prices.

[11] Detailed information of individual plants is shown in Canadian Wheat Board, (1992, p. 56). The data used here differ only slightly. The CWB data show an industry capacity in 1992 of 174 million bushels malt, whereas it is more readily accepted that the U.S industry is about 186 million bushels. The latter figures are used here.

[12] A minor share is held by a Japanese group.

[13] Brewers supply maltsters with barley and the maltsters collect a toll for processing services. This gives brewers complete control over barley selection.

[14] Industry estimates vary but generally suggest that Coors requires 100 percent 2-rowed, while Anheuser-Busch utilizes about 70 percent 6-rowed and 30 percent 2-rowed. Miller utilizes up to 20 percent 2-rowed, and other brewers utilize virtually all 6-rowed malts. Combining these figures with industry sales indicates that U.S. demand for 2-rowed malt is about 40 million bushels.

[15] Some Canadian plants reportedly are in the process of expanding due to anticipated exports to Asia and Japan or to increase efficiency.

[16] This is due to tradition. Europe dominates the world malt market and has trained brewers to use its 2-rowed varieties.

[17] For malt, the sum of initiatives include those that were canceled as well as those that did not result in transactions.

[18] This has accentuated pressure within Canada (i.e., from producers who are situated near the border) for dismantling of the Wheat Board's control over U.S. sales.

[19] Technically, the CWB has a monopoly on the procurement of barley for uses other than domestic feed, including sales of Canadian barley outside of Canada.

[20] See Wilson, 1993 for a discussion of these issues and implications in the case of wheat.

[21] The CWB has authority to issue export permits, which are required to legally ship barley to any other country, including the United States.

[22] In addition, a cap applies to the Canadian Government's share of total shipping costs.

[23] See Agriculture Canada (*Regulatory Review*, 1992) for a discussion of these issues.

14

American Feed Manufacturing Industry: Trends and Issues

Jack E. Houston

Trends in the Feed Industry

Feed grains, including corn, sorghum, barley, oats, and some wheat, account for over 25 percent of total crop production value in the United States, averaging $21.4 billion annually over the 1990–1994 period (Lin et al., 1995). Corn directly contributes more than 85 percent of the total value, and wet and dry milling of corn also produces high-protein grain by-products for the livestock feed industry. Weather phenomena and changes in government programs have been associated with larger-than-historical fluctuations in feed grain production, but the total disappearance has continued to trend upward. About 75 percent of the domestic use of feed grains is ascribed to livestock and poultry feeding during this period.

The emergence of on-farm feed mixing has jolted the feed supply industry, especially with large, integrated feedlots in cattle, hogs, and poultry (Kimle and Hayenga, 1993). The effects of this trend include increased concentration in the industry, with the smaller feed supply companies losing business and increasingly finding themselves competing against large national companies. Sizable livestock- and poultry-producing units can now mix and grind their own feed profitably, largely due to the increased size of their commercial production units and changes in technology. This trend is particularly evident in the poultry industry, which almost wholly has integrated vertically to include feed processing and distribution to associated producers.

Concentration of major players in the feed industry is one of the principal developments over the last two decades (Table 14.1). The number of prepared feeds firms (SIC 2048, with a 98 percent specialization ratio in primary product and excluding cat and dog feed) declined from 1,439 in 1977 to 1,160 in 1992, or 19.0 percent. Capital investment, primarily in technology, increased quite rapidly in the 1980s, although the rate of investment has declined somewhat in the 1990s. While all employment in the industry fell just 9.2 percent, employment in production activities declined 15.1 percent during the period 1977 to 1992.

Since the early 1980s, firms in the feed industry have been merging or acquiring other firms, their production capacities, and sales networks. These firms contend that they must take advantage of the abilities of larger size in order to more efficiently develop new products through research and development, to expand internationally, and to compete more favorably for market share in the United States. Archer Daniels Midland (ADM) has been especially adept at this strategy over the period from 1977 to present. They have acquired almost 30 percent of the soybean crushing capacity and 25 percent of the corn and cottonseed milling industries. In both of these industries, they were not even in the top four firms just 15 years ago. In 1994, ADM purchased the largest Canadian feed company, Maple Leaf. In late 1994, ADM and Ag Processing formed a 50-50 joint venture partnership, combining their feed manufacturing plants under Consolidated Nutrition, L.C. (*Feedstuffs*, 7 Nov. 1994). Consolidated Nutrition also combined Supersweet Feeds, Masterfeeds, and Master Mix Feeds, as well as purchasing all the feed mills that Central Soya operated. This new company is now the largest domestic feed manufacturing company.

Another development in the industry in 1994 saw Purina Mills acquire Ralston Purina's international feed manufacturing business, which effectively doubled the size of Purina Mills' capacity and sales (*Feedstuffs*, 21 Nov. 1994). While the top feed manufacturers may not control sub-

Table 14.1 • Prepared Feeds (2048), N.E.C. (animals and fowls, except dogs and cats; includes slaughter for feeds), Employment and Product Value, 1972–1992.

Year	Companies[a]	Plants[b]	All Employment	Production Employment	Product Value	Value Added
	number		*thousand employees*		*million dollars*	
1972	1,579	2,120	44.0	27.7	5,037.1	1,100.5
1977	1,439	2,063	39.1	23.2	8,786.8	1,544.7
1982	1,245	1,827	37.5	21.7	11,298.1	2,188.4c
1987	1,182	1,738	34.5	19.5	11,468.2	2,603.0
1992	1,160	1,714	35.5	19.7	14,373.9	2,875.6

Source: *Census of Manufactures.*

[a] Business organization consisting of one or more establishments under common ownership/control.

[b] Establishment with payroll at any time during year.

[c] 1982 change in reporting value of inventories.

stantial national market shares (Table 14.2), they often dominate regional or local markets (e.g., Agway in the Northeastern region, Farmland in the Plains and Midwestern regions, and Land O'Lakes in the upper Midwestern region). (Kimle and Hayenga, 1993, p. 25).

Increased diversification also has been a trend in recent years. Several feed companies have entered into animal and human medicine and research, developing feed for specialized markets, and generally becoming broader-based companies. Cargill expanded into production of ethanol gasoline additives from corn milling. Nationally, farmer co-ops have been consolidating along with the larger feed companies, seeking to increase sales and market share. Economic forecasters estimate that industry consolidation will continue, such that by the year 2000 there will be only 50 major companies worldwide in the agri-food sector (*Feedstuffs*,

Table 14.2 • Top Feed Manufacturers in United States (1989).

	Annual Capacity 1000 tons
1. Purina Mills (British Pet)	5,000
2. Wayne Feeds (Continental Grain)	2,500+
3. Goldkist	2,000
4. Nutrena Feeds (Cargill)	1,500+
5. Master Mix (Central Soya-Ferruzi)	1,500
6. Farmland	1,325
7. Agway	1,064
8. International Multifoods	1,001
9. Land O'Lakes	1,000
10. Moorman	1,000
11. Kent Foods	900+
12. Southern States Co-op	788
13. Hubbard Milling	700
14. Golden Sun Feeds	500
15. ACCO Feeds (Cargill)	450
16. GTA Feeds	400
17. Walnut Grove (WR Grace)	340
18. Country Mark	300
19. ConAgra Feeds	300
20. MFA	275

Source: Kimle & Hayenga (1991).

July 1994). Economies of scale and the ability of larger companies to open up new markets and develop new, higher-quality products will be the driving forces behind this trend.

Value added as a proportion of total shipped sales has fluctuated around 20 percent over the past two decades, ranging from 17.8 percent to 22.7 percent, with no clear trend during that time. Of the $14.37 billion shipped value of feed product in 1992, $2.88 billion (20.0 percent) was considered value added within the feed manufacturing industry (Table 14.3). Proportions of value added ranged from 8.9 percent in the chicken/ turkey feed sector to some 26 percent or more in the feed supplements/ concentrates products. While the dairy feed segments provided more employment than other feed types, the chicken/turkey sector invested $41.7 million in 1992, nearly double the next nearest feed manufacturing segment.

Related Feed Industry Markets

The American feed industry is highly integrated with other grain and oilseed processing industries through their by-product or multiple product input and output associations, including protein concentrates and other additives. Many national and multi-national corporations in the grain and oilseed sector envelop specialty feed processing and distribution firms. These linkages with the feed industry compel one to take a closer look at the major grain and oilseed processing firms, many of which are parent companies, and the changing industrial structure of their market environment.

Soybean Oil/Meal Crushing Industry

The employment and product value (*Census of Manufactures*, 1992) for soybean oil mills from 1972 to 1992 are shown in Table 14.4. Employment has been reduced by 18.7 percent over the 20 years. During this time, shipping value has nearly trebled, and increased by 41 percent over the 1977 figure. While the number of companies in the processing of soybeans decreased from 65 in 1977 to 52 in 1982, to 47 in 1987, and to 42 in 1992 (35 percent), plant numbers declined from 121 to 106 between 1977 and 1987, and further to 99 in 1992 (*Census of Manufactures*). Total crushing capacity declined also, from just over 127,000 tons/day in 1977 to 123,664 tons/day in 1982 and 108,159 in 1988 (Marion and Kim, 1991). Although total crushing capacity fell, industry concentration in-

Table 14.3 • Prepared Feeds (2048), N.E.C. (animals and fowls, except dogs and cats; includes slaughter for feeds), Employment and Product Value, 1992.a

Feed Type	Plantsb	All Employment	Production Employment	Product Value	Value Added	New Capital
	number	*thousand employees*			*million dollars*	
Chicken/turkey	201	5.3	3.2	4,938.4	439.4	41.7
Dairy	152	4.9	2.7	1,958.4	436.8	22.1
Dairy supplements/ concentrates	87	2.3	1.0	713.6	183.4	12.3
Swine	40	0.7	0.5	287.8	46.8	8.9
Swine supplements/ concentrates	183	6.1	2.7	2,063.9	542.8	19.7
Beef	65	2.1	1.3	588.9	127.4	10.6
Beef supplements/ concentrates	78	1.9	1.1	519.4	140.0	9.2
Total industry	1,714	35.5	19.7	14,373.9	2,875.6	

Source: *Census of Manufactures*.

[a] Not all feed types included.

[b] Establishment with payroll at any time during year.

Table 14.4 • Soybean Oil Crushing Industry (2075), Employment and Product Value, 1972–1992.

Year	Companies[a]	Plants[b]	All Employment	Production Employment	Product Value	Value Added
	number		*thousand employees*		*million dollars*	
1972	54	94	9.1	6.6	3,357.2	350.0
1977	65	121	9.4	6.6	7,580.0	373.8
1982	52	114	8.9	6.2	8,603.6	678.2[c]
1987	47	106	7.0	4.8	9,074.1	1,011.5
1992	42	99	7.4	5.1	10,650.6	1,273.8

Source: *Census of Manufactures.*

[a] Business organization consisting of one or more establishments under common ownership/control.

[b] Establishment with payroll at any time during year.

[c] 1982 change in reporting value of inventories.

creased from mergers and takeovers. The four largest firms in the soybean crushing industry from 1977 to 1988 are shown in Table 14.5.

Archer Daniels Midland (ADM) acquired four new plants in 1982, six in 1985, one in 1987, and their capacity nearly doubled (increased 72.3 percent) from 1977 to 1988. Meanwhile, Cargill acquired one plant in 1982, one in 1983, and five in 1985. Their crushing capacity increased 23.5 percent between 1977 and 1988. Likewise, Bunge acquired two plants in 1978, one in 1983, and two in 1984, increasing their crushing capacity 111.5 percent during the same period. Ag Processing acquired six plants in 1983, when it was formed as Co-op from the merger of Farmland, Inc., Boone Valley Co-op, and Land O'Lakes. Ag Processing acquired a further two plants in 1986. They had a 13,100 ton/day capacity in 1988 (Marion and Kim, 1991). The four-firm concentration of soybean crushing firms over the ten-year period in the data shows that market share for these firms increased by 64 percent (Table 14.5).

Wet Corn Milling Industry.

Employment and product statistics for the wet corn milling industry (*Census of Manufactures*) are shown in Table 14.6. After a dip from 26 to 22 companies between 1972 and 1977, the industry has demonstrated remarkable growth, increasing the number of companies by 27 percent between 1977 and 1992. While employment dropped some 15 percent, product value soared by 250 percent over the 15-year period. Value added in the

Table 14.5 • Largest Four Firms, Soybean Oil Crushing Industry, 1977–1988.

1977		1982		1988	
Firm	**Share %**	**Firm**	**Share %**	**Firm**	**Share %**
Cargill	15.1	ADM	17.8	ADM	29.9
ADM	14.6	Cargill	15.8	Cargill	21.9
Central Soya	9.9	Central Soya	9.7	Bungee	12.8
Ralston Purina	6.8	Bungee	7.6	AgProcessing	12.1
Total Four-Firm	46.4		50.9		76.4

Source: Marion & Kim (1991).

Table 14.6 • Wet Corn Milling Industry (2046), Employment and Product Value, 1972–1992.

Year	Companies[a]	Plants[b]	All Employment	Production Employment	Product Value	Value Added
	number		*thousand employees*		*million dollars*	
1972	26	41	12.1	8.4	832.3	331.2
1977	22	39	10.9	7.8	2,014.8	666.7
1982	25	42	9.5	6.7	3,268.4	1,157.4c
1987	31	60	8.6	5.9	4,788.9	2,074.5
1992	28	51	9.2	6.1	7,045.2	3,257.5

Source: *Census of Manufactures.*

[a] Business organization consisting of one or more establishments under common ownership/control.

[b] Establishment with payroll at any time during year.

[c] 1982 change in reporting value of inventories.

industry totaled nearly $3.26 billion in 1992 on total shipping value of $7.05 billion.

The wet corn milling industry (SIC 2046) increased in concentration due to acquisitions and the increased capacity of the large firms in the industry (Table 14.7). Plant numbers rose from 19 in 1977 to 23 in 1988, while capacity increased from 1.34 million bushels (bu.) in 1977 to 2.07 million bu in 1982 and 2.34 million bu in 1988 (Marion and Kim, 1991). The four-firm concentration share in the corn milling industry increased by 13.8 percent over the period 1977 to 1988.

Cottonseed Oil Milling Industry

Plant numbers in the cottonseed oil milling industry (SIC 2074) fell from 115 in 1972 to 77 in 1982 and to just 45 in 1992, while the number of companies declined from 74 to 22 between 1972 and 1992 (Table 14.8). Milling capacity declined from 20,505 tons/day in 1977 to 14,405 tons/ day in 1988 (Marion and Kim, 1991), but total product value has tended to fluctuate greatly over this period.

The cottonseed milling industry historically has had a low concentration level, but from 1977 through 1988, the four-firm concentration share increased by 38 percent (Table 14.9). ADM made several acquisitions during this period to increase their capacity from 1,135 tons/day in 1977 (not a top four firm) to 3,235 tons/day in 1982 and to 3,585 tons/day in 1988, garnering 24.9 percent of market share in 1988. Anderson Clayton was acquired by Quaker Oats in 1986, but Paymaster Oil Mill and Western Cotton Service were the two major parts of its cottonseed operations and were divested separately. Co-operatives, such as Plains Cooperative Oil Mill with 8.3 percent of 1988 industry capacity, held a large and increasing share of the total industry capacity: 37 percent in 1977, over 40 percent in the 1980s, and 45 percent in 1988 (Marion and Kim, 1991).

Concluding Industry Observations—Trends and Issues

Of the three intermediate industries closely related to the feed industry, the corn milling industry demonstrated the highest growth over the period 1977 to 1994. Employment in this industry was little affected, partially because of increased numbers of firms and capacity. Total sales swelled by 250 percent over 1977 industry figures, very high growth for any industry. Soybeans continued to demonstrate strong growth in sales, while employment declined moderately through capitalization and plant

Table 14.7 • Largest Four Firms, Corn Milling Industry, 1977–1988.

1977			1982			1988		
Firm	Share	Capacity	Firm	Share	Capacity	Firm	Share	Capacity
	%	000 bu/day		%	000 bu/day		%	000 bu/day
CPC	18.9	254	ADM	26.1	540	ADM	27.8	650
AE Staley	18.6	250	AE Staley	18.3	380	AE Staley	20.1	470
Cargill	14.9	200	CPC	17.9		Cargill	16.9	395
Clinton Corn	10.4		Cargill	14.5	300	CPC	13.4	312
ADM	5.9	80						85.9
Total four-firm	62.8			76.8			78.2	

Sources: Marion & Kim (1991): *Census of Manufactures*

Table 14.8 • Cottonseed Oil Industry (2074), Employment and Product Value, 1972–1992.

Year	Companies[a]	Plants[b]	All Employment	All Production	Product Value	Value Added
	number		*thousand employees*		*million dollars*	
1972	74	115	5.5	4.4	458.7	88.2
1977	62	97	5.2	4.1	859.2	197.4
1982	47	77	5.2	4.3	933.3	202.9c
1987	31	52	2.6	2.0	470.7	106.9
1992	22	45	2.4	1.9	737.8	211.4

Source: *Census of Manufactures.*

[a] Business organization consisting of one or more establishments under common ownership/control.

[b] Establishment with payroll at any time during year.

[c] 1982 change in reporting value of inventories.

Table 14.9 • Largest Four Firms, Cottonseed Milling Industry, 1977–1988.

1977		1982		1988	
Firm	**Share %**	**Firm**	**Share %**	**Firm**	**Share %**
Anderson Clayton	10.6	ADM	17.6	ADM	25.9
Buckeye Celulose	9.2	Anderson	11.8	Plains	8.3
Ranchers Cotton Oil	7.8	Ranchers	8.7	Paymaster	6.9
Plains	5.9	Plains	6.5	Western Cotton	5.2
Total Four-Firm	33.5		44.6		46.3

Source: Marion & Kim (1991).

scale. The 35 percent decrease in number of companies demonstrates sharply increasing concentration in the soybean crushing sector. The cottonseed oil industry fell on hard times in the late 1980s, but it may now be rebounding somewhat. Concentration of the industry is exhibited in the change in the number of companies (one-third of 1977 number), and employment has been off by half of 1977 levels. The fact that sales fell only 14 percent during this period suggests that while consolidation has occurred, the volume of through-put has fallen to the larger firms.

Current issues pressing participants in the prepared feed industry include the public relations questions raised by growth promotants and other feed additives. Not only do the firms face increasing consumer concern stemming from the influence of the use of these feed additives on final products, such as beef, pork, milk, and poultry, but they also face various liability questions and new government regulations. Like their cohorts in the food industry (including some of the same parent companies), feed labeling regulations require tracking of inputs, additional cost, and public scrutiny, including additional liability questions.

Expanding trade in feeds and feedstuffs, especially with Mexico, is also on the table for the more aggressive companies following NAFTA.

In particular, corn and corn product derivatives may be stimulated by increased livestock and poultry finishing in Mexico. However, perhaps the greatest recent issue touched off as a result of the changes in firm structure and sizes in the feed industry has been a price-fixing/anti-trust probe by the U.S. Justice Department. Four feed companies, including Archer Daniels Midland, A. E. Staley, Ralston Purina, and Cargill, have been under investigation, publicly since mid-1995. The outcome of this antitrust investigation may have far-reaching implications for the industries associated with feeds.

References

Kimle, Kevin L., and Marvin L. Hayenga. "Structural Change among Agricultural Input Industries." *Agribusiness: An International Journal.* 9(1993):15-28.

Lin, William, Peter Riley, and Sam Evans. Feed Grains: Background for 1995 Farm Legislation. U.S. Department of Agriculture, Economic Research Service, Agricultural Economic Report Number 714. Washington D.C., 1995.

Marion, Bruce W., and Donghwan Kim. "Concentration Change in Selected Food Industries: The Influence of Mergers vs Internal Growth." *Agribusiness: An International Journal.* 7(1991):415-432.

U.S. Department of Agriculture, *Feedstuffs Magazine.*, Washington. D.C., various issues.

U.S. Dept. of Commerce, Bureau of the Census. *1992, 1982 Census of Manufactures.* Washington, D.C.

15

Summary and Conclusions

Bob F. Jones and Robert L. Oehrtman[1]

The Turning Point

A series of events in the first half of the decade of the 1970s represent an important turning point in the structure of the grain marketing system in the United States. The sudden and unanticipated increase in U.S. grain exports set off a chain of events that reverberated throughout the agricultural economy.

The agricultural commodity export bubble continued to expand throughout the 1970s. Although it did not burst in the early 1980s, it did begin to shrink, especially for the United States. During the expansion phase, the grain market system quickly expanded its capacity to handle larger exports. Adjusting later to smaller export volume was much more difficult. Both the expansion and contraction phases intensified significant structural change in the system.

---

[1] Helpful comments on an earlier draft were provided by Paul Farris, Professor Emeritus, Purdue University.

This expansion in demand for U.S. grains contrasted sharply with experience over most of the previous 40 years. From the beginning of the 1930s, the agricultural economy was characterized by slow growth in domestic demand for agricultural products, very small grain exports, and advances in production technology that generated increasing supplies of grain with relatively low farm prices and farm incomes. National agricultural policies were designed to place some limits on grain production increases in order to support farm price levels and incomes to farmers. The intention was to increase per capita farm income, because it was significantly lower than the income of people in the nonfarm sector and to facilitate adjustments to changing technology and resource use in agriculture.

During this period, the agricultural sector was characterized as having excess capacity and the need to shift resources (especially labor) out of the sector to the general economy where resource returns and incomes were higher. This long period of excess capacity and low farm income was interrupted by two brief but important intervals, 1941-1946 and 1951-53 when demand was driven by war-time needs.

Coinciding with excess capacity in grain production, the grain marketing system had evolved into a system dominated by local elevators, many of which were farmer-owned cooperatives that assembled grain from farmers, then shipped the grain to central markets for processing or redistribution to processing centers. Rail shipment was the principal means of transport, with trucks used primarily for short hauls. A large proportion of feed grains was fed to livestock and poultry on farms where the grain was produced. Then livestock were shipped to central markets for slaughter with meat products shipped on to consumption centers.

Throughout the 1930-1970 period the federal government was indirectly involved in the grain marketing system through production adjustment on farms and directly in grain storage programs. Grain storage programs featured government ownership of grain storage facilities in areas where grain was produced with facilities to be used for storage of grain acquired by the Commodity Credit Corporation (CCC) and subsidization of the cost of facilities for farmer storage of their own grain. A significant proportion of the annual harvest of grain was put under loan to the CCC and later turned over to it through the non-recourse loan program.

Up until 1972, except for the two war-time periods, grain prices tended to be relatively stable. As government grain storage programs grew in importance, especially after the Korean Conflict, an increasing proportion of carry-over grain stocks was under government control. Prices tended to stabilize around price support levels. Release price rules allowed only small price changes in response to annual changes in demand and production so long as the CCC held large stocks of grain. Consequently, the grain marketing system became accustomed to relatively stable prices and dependence on storage income for a significant share of its total income.

As noted above in this chapter, the sudden and unanticipated increase in United States grain exports starting in 1972 was a shock that is still having an impact on the grain marketing system. As noted in Chapter 1, the grain marketing system in the United States is a dynamic system that responds to market forces. Although some of the forces identified here have their roots in earlier decades, the 1970s saw these forces come together in such a way as to have dramatic impacts on the system.

After the turning point, export demand for grains and oilseeds was the dominant growth area for agricultural products. Domestic demand continued to grow at a slow rate, but was changing due to changes in composition of the diet and demographic shifts in the composition and location of population.

Changes in technology at the farm level created new challenges for the grain marketing system. In particular, changes in corn production created significant challenges to the marketing system. The combination of yield increases, changes in harvesting methods and development and use of mechanical grain dryers resulted in significantly larger quantities of corn requiring storage space and transportation in a much shorter period of time than in earlier decades.

As more grain became available in a shorter harvest period with the need to ship it to export points, railroad transport underwent a revolution. Railroad deregulation and use of multiple-car rates facilitated the movement to larger shipments and larger grain storage facilities generally away from traditional central market locations. As new destinations needed to be served, truck transportation over longer distances, and barge transportation handled larger quantities of grain.

As government-owned grain stocks were sold (1972-75) and annual carryover stocks became much smaller, within-year grain price changes became much larger. As less grain moved through central markets, new pricing procedures needed to be established. With prices becoming more variable, the futures market, forward pricing and use of options assumed a greater role as participants in the system found ways to deal with risk emanating from demand, supply and price variability.

In 1972 United States grain exports jumped from 37 million to 66 million tons, up almost 77 percent from the previous year. Subsequently, grain exports continued to grow, reaching a peak of 114 million tons in 1980. The U.S. share of the world grain market increased from 34 percent to 42 percent in one year. Soybean exports increased from 13 million to 14.7 million tons. However, with rapid growth in exports of Brazilian soybeans, the U.S. share slipped from 85 percent in 1972 to 81 percent of world exports in 1973. By 1981, the peak year for U.S. soybean trade, exports had reached the 25 million ton level.

The export boom from 1972 to 1980 was driven by a decline in world grain production (1972-74), a shift in U.S. exchange rate policy, continued world economic growth, a dramatic increase in petroleum prices, and trade and price policies followed in centrally planned economies. Although there is agreement that each of these developments contributed to the boom, there is not complete agreement on the relative weights to assign to each development.

The immediate U.S. response to this surge in demand was to sell grains already in storage. By the end of 1974, the CCC had sold all of its stocks of wheat. From the early 1950s through 1972, 50 to 60 million crop acres were held out of production (idled) under government programs. By the 1974 and 1975 crop years no land was idled under government programs. Additional land in production and additional demand for grain led to massive investment by farmers in new, larger machinery, buildings, and storage facilities. After the CCC had disposed of its grain stocks, it divested itself of government-owned storage facilities. The grain marketing system began a storage expansion phase, investing in new, larger elevators and acquiring, through lease or ownership, fleets of railroad cars for transport of grain to export points. Off-farm grain storage capacity increased from 5.7 million bushels in 1972 to 7.2 million bushels in 1980 when U.S. exports reached their peak volume.

For a short time, the government was out of the grain storage business. However, with rising production costs, and passage of the 1977 Agricultural Act that established higher loan rates and the Farmer-Owned Reserve (FOR), the CCC was soon back in the business of acquiring grain from loan takeovers or of subsidizing grain storage in the FOR.

U.S. grain export growth from 1972 to 1980 was followed by a sharp and continued decline through 1985. One major reason for the decline was U.S. monetary policy used to control inflation. That policy contributed to exchange rate adjustments in the value of the U.S. dollar and worldwide recession. The 1980 U.S. grain embargo of shipments to the Soviet Union was another unsettling factor. In addition, the U.S. policy response as contained in the 1981 Agriculture Act, the collapse of Eastern European demand, and increased grain production in the European Community (now called the European Union), are all cited as developments that contributed to declining U.S. grain and oilseed exports.

Having gone through a decade of increasing grain exports and rising production costs, the Agriculture and Food Act of 1981 was designed to protect producers at a new higher price level with higher loan rates and larger direct payment to producers. The ink was barely dry on the Act when the economic environment for agriculture changed. As grain exports declined sharply and inflation was brought under control, provisions of the Act were encouraging producers to put grain into government storage or the FOR rather than into the export market. It was not until the 1985 Act was passed that a policy was implemented which would stimulate exports rather than long-term storage of grain.

Changes in demand placed upon the grain marketing system resulting from changes in economic variables such as grain production, exports, transportation, and government programs are frequently abrupt and difficult to predict. Hence, investments in infrastructure are often risky and sometimes result in large losses. The system can move from under capacity to excess capacity in a short time span. This induces structural change in the system.

Structural changes in the U.S. grain marketing system have been more extensive and more comprehensive in recent years than in any time in history. The late 1980s and early 1990s can best be characterized as a period of consolidation and increased concentration in grain marketing. Before looking more specifically at those changes it may be helpful to briefly summarize some of the economic theory affecting structure and expected implications.

Economic Theory Affecting Structure with Implications

In Chapter 4, Ladd presented conventional economic theory that pertains to economies of scale in grain handling facilities and the implications for changes in the structure of the grain marketing system. A principal focus was to relate the theory to technological and institutional developments in the transportation system. Those changes included replacement of boxcars by covered hopper cars, introduction of unit trains and unit train rates, rail line abandonment, and rail deregulation. He also observed the nature of capital markets, which exhibited economies of scale in their ability to raise capital to finance technological advances. His conclusion was that the nature of those technological developments, the existence of economies of scale, and the nature of capital markets taken together favored larger firms. Consequently, smaller firms were at a competitive disadvantage, and there was economic pressure to enlarge facilities, merge with other smaller firms, or go out of business.

Technological developments at the farm level enabled producers to harvest a much larger volume of grain in a shorter period of time. As export demand grew and unit train freight rates came on the scene, demand grew for larger-scale elevators within the producing areas which could store sufficient quantities of grain to fill a unit train. This development led to growth of subterminal elevators in producing areas that could ship large volumes of grain to export elevators or directly to processing centers.

Ladd stated that a number of observers identified the existence of excess capacity in the U.S. grain marketing system that followed the increase in investment in grain handling facilities and the subsequent decline in export demand. He concluded that excess capacity was inevitable and desirable in an economy where there existed: (1) limited capacity and sharply rising short-run average cost curves with higher volume, (2) large and unpredictable variations in volumes of farm production, marketing, and exports; and (3) declining unit capital costs as capacity is expanded.

An overview of several studies of rail-based grain distribution systems examined the consequences of rail deregulation, mainly the Staggers Act. The Staggers Act of 1980 granted railroads the freedom to establish freight rates and provided various means of making upward rate adjustments, permitted them to enter into contracts with shippers and receivers, and allowed contract rates based on seasonal or other demand conditions. These provisions virtually eliminated the legal collusive rate-setting power of rate bureaus.

Structural Change in the Grain Marketing System

Country elevators have become larger as local grain marketing cooperatives have merged to achieve lower costs associated with larger volume. Other smaller country elevators have gone out of business. As noted, unit trains and multi-car freight rates have stimulated the development of subterminals. Some are new facilities, but others are enlarged country elevators. Subterminal elevators usually ship grain to export elevators or domestic processors and end users, not to central markets as was the case two or three decades ago.

One consequence of the changes in routing of grain was the decline in cash trading of grain at grain exchanges. Futures trading evolved out of cash grain trading at grain exchanges and the importance of futures trading has increased over the years. Buying and selling of grain based on visual inspection of a sample has been replaced by forward "to arrive" cash contracts between country elevators and grain merchants where price, grade, premiums, and discounts for quality are agreed to in the contract. Today, most cash grain is traded by telephone. With the increased roles of subterminals and shipping grain in unit trains, the system has become more decentralized. Futures prices have become even more important as the "basis" for pricing cash grain.

As futures prices came to have a greater role in the pricing of grain, the volume of contracts traded on grain exchange expanded significantly. In 1980, 39.5 million contracts were traded. Corn alone had a volume of almost 12 million contracts, representing almost 60 billion bushels of corn. This was approximately ten times the volume of cash corn that was actually traded in that year, indicating significant within-year evaluation and reevaluation of changing market conditions, as well as speculation in the market.

While grain trading has become more decentralized with the decline of central markets, ownership and operation of facilities have become more centralized. When one compares the ten largest multiple facility grain companies by number of facilities and capacity in 1995 with the ten largest in the number of elevators and capacities for 1981, one of the striking features is the new names that appear in 1995 and the shifting of position among the top 10 companies. However, Cargill maintained its position as the largest grain company. Firms have become larger, primarily through acquisition of other firms.

Structural changes in grain marketing cooperatives have occurred at both the regional and local levels. In 1981, there were 16 regional and three interregional cooperatives that handled 3 billion bushels of grain. Through a process of mergers with other cooperatives, or sale of assets to other firms, these 19 firms became four regional cooperatives by 1995. Regional cooperatives are federations of locals. The number of local cooperatives in the United States declined from 1,623 in 1985 to 1,193 in 1993.

Major multinational cooperatives are large firms that operate globally and handle much of the grain that is bought and sold in the world today. However, they are not the only exporters. As noted previously, farmer-owned cooperatives downsized grain operations in the mid-1980s but have expanded direct grain exporting in the 1990s.

Joint ventures are new structural innovations in the grain industry and can be classified as: (1) joint ventures between regional grain cooperatives and investor-oriented firms (IOFs); and (2) joint ventures between IOFs. Several have been dissolved after serving as a preliminary step toward full acquisition by one of the partners.

Vertical integration is proceeding rapidly in the grain industry. Most of the new investments in the industry are in value-added grain processing activities by farmer-owned cooperatives. These developments appear to be a response to: (1) narrow profit margins in traditional grain merchandising activities because of excess capacity in the system; and (2) a search for new profit opportunities believed to exist in value-added products.

Futures, Options, and Cash Grain Prices

There have been dramatic changes in the marketing system for price discovery and risk management in the United States grains sector during the past 20 years. This is reflected in the rapid escalation in the volume of futures trading of corn and soybean contracts at the Chicago Board of Trade (CBOT) and the introduction of options trading for agricultural commodities.

The expansion in futures trading has not been limited just to grain. The volume of trading in livestock futures and other futures contracts including metals, financial instruments ,and market indexes from 1980 to 1997, have also increased dramatically. However, changes have varied by sector with the greatest growth occurring in the financial, metal, and index futures.

The shipping, storage, and the pricing of cash grain have become more decentralized with the decline of the traditional central markets in Chicago, Kansas City, and Minneapolis. This decentralization along with increased need for risk management has led to the increased use of futures markets and risk management techniques.

The challenges of risk management have become much greater as the export market, with its inherent variability, has become a much larger part of the market for U.S. grains and as government programs have changed over the last 20 years. Changes in farm programs starting with the 1985 Farm Bill and modified by the Freedom to Farm Act of 1996 have drastically reduced the pricing safety net available to farmers and to some extent, grain merchandisers, and have effectively exposed them to increased market risk.

In response to the demand for more risk management tools, the use of options grew as they became part of different risk management strategies. The new forward contracts that have been developed by grain buyers permit individuals to take advantage of options indirectly through merchants without being directly involved in the options market or needing to understand the mechanics themselves.

The authors conclude that the country elevator is in an ideal position to offer market pricing tools to farmers and to use this array of tools as a way to compete for grain. Those elevators that are most innovative in their offering of contracts to farmers stand a greater chance of acquiring most of the farmers' business.

One advantage of these new hybrid cash grain contracts from the producer's perspective is that it removes the burden of having to be responsible for margin requirements. In addition, the elevators may have an advantage in acquiring margin financing because of their ability to aggregate positions and realize economy of size effeciencies that are present in using futures contracts. Also, from a producer's perspective, the contract may allow deliveries of odd lot sizes rather than the standardized 5,000 or 1,000 bushels required per grain futures contract.

The U.S. futures industry is in a state of transition as it responds to many structural changes in the U.S. and global economies, however, the issue of delivery mechanisms and delivery points which grew as central markets declined continues to generate problems. These problems are also related to changes in rail rates which followed passage of the Staggers Act of 1980 that tended to increase freight rates for processed products versus raw commodities. This forced the relocation of the processing industries closer to the demand centers for processed products, resulting in a more geographically dispersed food processing industry.

Adding more delivery points to increase deliverable supplies has been one approach to this problem. However, adding delivery points leads to other problems. Delivery continues to be an evolving area with some problems yet to be worked out.

The "open outcry" system of establishing prices on the futures trading floor is being challenged by development of electronic trading of commodities. To date, electronic trading is in use for grains on a limited basis. Its use has been limited in part because of the complexity of some trades. Almost everyone in the U.S. futures industry will accede to the opinion that eventually all of the exchanges must go electronic. The main point of debate is how quickly this change should be implemented and what path it should take toward full implementation.

Grain Grades and Quality

Grades and standards provide a facilitating function in grain marketing as they increase the efficiency and effectiveness of marketing transactions. Grades and their associated descriptive terminology provide information to enable buyers and sellers to identify value without visual inspection (at all points of sale).

Factors to be measured and factor limits for each grade have changed many times in the history of grades. The role of grades has become more important in grain trade as the system has become more decentralized and a larger proportion of grain and oilseeds has gone to the export market.

Grades have often been a controversial attribute of grain trading. Controversy arises over consistent application of criteria in setting grades and occasional outright fraud, a case at the ports in 1975 being the most recent notable example. The scandal at the ports resulted in revisions and reorganization of the U.S. system of grain grades and standards. Following this scandal, Congress created the Federal Grain Inspection Service (FGIS). This agency was given responsibility for research and development related to grades and standards, implementation and enforcement of grade-related regulations and supervision of the licensed inspection agencies. The amendment required that a federal employee grade all grain for export. Because of this action, the federal government was given a larger role in restoring credibility to the system.

However, these changes did not eliminate all concerns about U.S. grain grades. As grain exports declined in the early 1980s, the search for causal factors identified U.S. grain grades as a contributing factor. This assertion set off even more inquiry and research to evaluate the existing grain grading system.

The official grades and standards provide international traders with a language enabling grain to be purchased from U.S. origins with confidence that the quality loaded on vessels meets the specifications in the grades. However, many buyers and processors are interested in information not provided by the grades, e.g., oil content of soybeans and protein content of wheat. Consequently, processors are increasingly using contracts that specify quality attributes including variety, cultural practices, harvesting methods, and storage and drying technology. The use of contracts that specify more attributes will likely continue to grow as technology for measuring attributes becomes more sophisticated and enables the generation of readings more quickly.

Grades and standards are a public good. Inspection by an objective third party agency with no interest in the transaction is essential to credibility. Beneficiaries of uniform grades include farmers, marketing firms, processors, consumers, exporters, and foreign buyers. However, the third party attribute has come under increasing pressure. User fees paid to the federal government have become more common because of national policy actions. These user fees place the primary financial burden on exporters by virtue of the mandatory export inspections; domestic firms can and do shift to house grades and independent agencies and laboratories. User fees that move the U.S. grain inspection system back toward the chaos of the 1980s will likely add cost to the transactions, decrease efficiency, decrease our attractiveness to foreign buyers, and further increase inequities among marketing firms of different sizes.

Sanitary and phytosanitary (SPS) technical barriers are increasingly recognized as important grain marketing hurdles. These barriers may be disguised mechanisms designed to protect domestic producers from foreign competition. In other circumstances, animal and plant health issues catalyze restrictions for legitimate purposes. The challenge is to find a technically and economically feasible way to resolve the problem. Chapter 7 presents a study of wheat trade where exports have been restricted because of a quarantine on wheat suspected of being contaminated by TCK smut. The study presents an economic evaluation of possible ways to remove a quarantine imposed by China on wheat shipments through

Pacific Northwest ports, by shipping TCK-free wheat through identity preserved channels. A technology-cost assessment study was made of alternative ways to ship the wheat followed by an analysis of the potential benefits and costs of removing the quarantine.

The TCK quarantine imposed by China on Pacific Northwest wheat exports has been in existence for 25 years. The study concluded that there is little assurance that TCK smut can be economically eradicated to meet a zero tolerance standard or be isolated by identity preserved transportation and distribution systems. Further advances in technology could conceivably break through economic barriers and eliminate or isolate dwarf bunt, but costs would have to fall radically. Focusing on procedures to eliminate or destroy viable TCK spores at the export elevator spout might be a possibility with some modification of the present zero tolerance standard. Also, effective and low-cost treatments applied at the time the cargo is unloaded in a Chinese port might have potential. However, if China maintains its current zero tolerance standard against dwarf bunt spores in wheat shipments from the Pacific Northwest, options are extremely limited.

Since a quick cheap technological fix is not in the offering at this time, the industry may have to live with the quarantine indefinitely. However, there may be a basis for doubt that wheat shipments from the U.S. Pacific Northwest are a legitimate threat to China's domestic wheat production. A reexamination of the rational for the quarantine may be the best prospect for removing the marketing barrier. If it can be established, by thorough review and mutual resolution by scientists in the international community, that dwarf bunt infestation from wheat shipped through Pacific Northwest ports does not present a threat to China's domestic wheat production, then international dispute resolution measures might provide the channel to remove the quarantine.

Although this study covers only one product to one country, it concerns an issue that may become more important. As tariffs and quotas are reduced worldwide through various rounds of trade negotiations, domestic producers and other groups within the importing country may seek alternative methods for protecting domestic markets. Use of pesticides, various feed additives for producing livestock products, and bio-engineered products become target areas where trade disputes arise and lead to restrictions on trade.

Structural Change in the North American Flour Milling Industry

For several decades before the 1980s, demand for wheat flour products in the United States had minimal to negative growth. Since the early 1980s, however, demand for milling has increased because of a reversal of per capita consumption trends that have increased 3 to 5 percent per year. This growth in consumption is in part due to the development of new wheat products and growth of the fast food industry. Changes in transportation technology and freight rates as discussed previously affected the location of milling facilities. Discontinuance of the transit privilege and use of unit trains, which facilitate shipment of wheat but not flour, favored transition of the milling industry away from traditional milling centers toward destination markets. As firms adjusted to these changes, a combination of firms exiting and mergers contributed to consolidation in the flour milling industry. This consolidation has taken place across borders of the United States, Canada, and Mexico.

The flour milling industries had evolved independently in the United States, Canada, and Mexico. Policies, regulations, and the competitive structure of the vertical marketing system differed drastically across these countries. Exports between countries had been virtually non existant, and plants generally ship within their national boundaries without being impacted directly by developments in contiguous countries. This was primarily a result of trade policies followed in the three countries. However, features of the Canadian-United States Trade Agreement (CUSTA) and the North American Free Trade Agreement (NAFTA) are having important impacts on the structure of the industry.

The flour milling industry had four important developments: (1) Although the number of firms and plants has decreased, plant and firm capacity has increased. The numbers of multi-plant firms and firms in the "grain" group have increased; (2) The geographic boundaries among these firms are becomming blurred. However, each is gaining dominance in particular regions; (3) Procurement is becoming a major element of strategy for both Canadian and Mexican mills, an area previously protected because of their agricultural and trade policies; and (4) The extent of integration varies in these countries: U.S. firms largely integrated backward while Canadian and Mexican firms integrated forward.

Growth and Structure of the Wet Corn Milling Industry

The wet corn milling industry has grown rapidly in recent times; processing 1.3 billion bushels of corn in 1995 compared with 262 million bushels in 1972. This growth reflects large increases in demand for two major products, high fructose corn syrup (HFCS) and fuel alcohol. HFCS has replaced most other caloric sweeteners in the nonalcoholic beverage market. Fuel alcohol demand was stimulated by crude oil shortages in the 1970s and by the 1977 Clean Air Act with 1990 amendments. HFCS and fuel alcohol production accounted for over half of all corn used in wet milling production in 1993-95. Demand has also expanded for a wide range of other processed corn products used in food, paper, building materials, and textiles.

Improvements in the highly technical production process and plant scale economies reduced the number of workers employed in wet corn milling from 12.1 to 9.2 thousand between 1972 and 1992, even though industry capacity approximately tripled. Only four firms account for the bulk of industry output, representing a concentration of 73 percent of production in 1992. Market structure and performance issues have arisen within the industry over time, the most recent involving a price-fixing conspiracy for lysine in the early 1990s.

While the dramatic growth in wet corn milling moderated in the late 1980s and 1990s, further increases will likely exceed U.S. population growth due to technological advances and new product development. Foreign investment by U.S. firms is expected to grow, and exports of processed corn milling food products seem likely to rise. The total number of corn refining firms in the United States is not expected to change greatly, although further investment in corn refining capacity within existing firms can be expected with increased future demand in starch both for food and industrial uses.

Soybean Processing Industry: Structure and Ownership Changes

U.S. soybean processing and grain industry structure and ownership have changed dramatically through time. Factors affecting structure include supplies of soybeans, competition from other producing countries, and changes in domestic and foreign demand for soybeans and soy products. Structure is also affected by changes in government policies and programs directed to agribusinesses, farmers, and exporters. During the

1980s the supply of soybeans and the amount of soybeans crushed increased. Domestic demand for soybean meal and oil increased. On the world market, U.S. exports of soybeans and soy products encountered increasing competition from Argentina and Brazil in particular. Market shares of these countries increased while U.S. shares declined.

The four largest firms' share of soybean processing capacity increased from 46 percent in 1977 to 83 percent in 1995. While both the number of firms and the number of plants decreased, average crushing capacity per plant and per firm increased rapidly. Mergers, acquisitions, and consolidations explain the dramatic change in structure. Much of the structural change appears to be motivated by excess capacity and small operating margins in grain merchandising. Firms searched for new business opportunities in more profitable value-added businesses.

Major grain trading companies also became major soybean processors in the 1980s. In many firms, these mergers and acquisitions impacted grain trading as companies saw opportunities to vertically integrate their operations. For example, Cargill reorganized to integrate grain origination, soybean processing, and livestock feeding to handle more grain in-house. Operations of this type do not buy and sell grain but use transfer pricing to move grain among profit centers. If this trend continues and is used more by other firms, a consequence may be less competitive grain markets that lead to anti-competitive behavior in the buying and selling of grain. Consumers and farmers may see the effect of these trading practices in terms of larger operating margins, lower farm prices, and higher consumer prices.

Structural Changes in the U.S. Oilseeds Processing Industry

Soy oil competes with cottonseed oil, palm oil, rapeseed/canola oil, and peanut oil, along with several other oils of less importance. Since the early 1980s, soybean oil has lost part of its market share to canola, cottonseed, and peanut oils.

Changing attitudes toward foods that are perceived as beneficial or harmful to human health have become a major force in shaping consumption patterns in the United States. Through the 1970s, the edible fats and oils sector had been largely characterized by a high degree of commodity interchangeability, due to apparent minor differences in the physical and chemical characteristics of these oils. Commodity promotion and product advertising stimulated major changes in consumer attitudes toward edible fats and oils in the 1980s. Health concerns over cholesterol levels became common. These changes have, in turn, induced new or modified products and processes and encouraged industrial behavior and structural changes. Three events sparked public comment and significant changes in the edible oils industry sector: certification of edible rapeseed/canola oil development; a promotional campaign raising health issues associated with the consumption of tropical oils; and an emergency Chicago Board of Trade action in soybean futures.

While the promotional campaign against tropical oils served to educate consumers about the characteristics of various oils, it may have had an impact not foreseen by its sponsors. Negative publicity about one or more competitive products can raise questions about relative health and safety factors for all products that are closely related. Embarrassingly, this may boost another competitor's products more than one's own. Trends in the late 1980s and early 1990s suggest that the campaign against tropical oils benefitted corn oil and the newly introduced canola oil at the expense of soybean oil market share.

U.S. Rice Milling Industry: Structure and Ownership Changes

Any study of U.S. rice is unique among grain crops because its production and related value-added activities occur in only six states: California and the five contiguous southern states of Texas, Louisiana, Mississippi, Arkansas, and Missouri. Chapter 12 offers information about the changing structure and ownership of the highly concentrated U.S. rice milling industry between 1972 and 1994. That information provides a basis for considering the influence that selected changes in structure and ownership might have upon future performance in the rice milling industry.

In 1972, the U.S. rice milling industry was structured to serve a market that was 60 percent export and 40 percent domestically oriented. Despite the fact that U.S. rice production, as a percentage of world production, was only a low single digit number, the United States was the second largest exporter of rice in the world. Its exports were differentiated by high quality and varied type. The number of rice mills increased dramatically in Arkansas, Mississippi, Texas, and California between 1972 and 1985 as a consequence of the elimination of acreage control in 1974 and a 1973 expansion in exports. The decline in export demand resulted in excess milling capacity that was eliminated during the early 1990s through the closure of medium size export dependent mills primarily in the southern Gulf States of Louisiana and Texas. The decline in export demand was accompanied by increases in domestic demand that brought about a milling industry with a 45 percent export and 55 percent domestic orientation. The consequences of the change in market orientation included a shift in milling from the Gulf Coast states of Louisiana and Texas, to Arkansas and California, along with a decline in the 60 percent concentration of milling capacity held by the top five mills.

Studies of mill sizes and average costs between single floor and multi-floor mills that package both full line and 70 percent bagged rice across the six states, document the higher costs of Texas and Louisiana mills. These studies show the significance of volume in lowering the average costs of milling particularly with single-floor as contrasted with multi-floor mills. Consideration of milling margins in California relative to milling margins in the contiguous southern states suggests the effects that competition and union labor have upon milling margins. Milling margins were found to be sensitive to changes in rough rice and f.o.b. mill prices.

The American Feed Manufacturing Industry

Feed grains account for over 25 percent of total crop production value in the United States. Weather phenomena and changes in government programs over the past two decades have been associated with larger-than-historical fluctuations in feed grain production, but the total disappearance has trended upward. Emergence of on-farm feed mixing has shaken the feed supply industry, especially with the advent of large, integrated feedlots and poultry operations. The effects of this trend include increased concentration in the industry. The number of prepared feeds firms declined 19 percent from 1977 to 1992, while employment in production activities declined 15.1 percent during this period. Capital investment, primarily in technology, increased quite rapidly in the 1980s, although the rate of investment declined somewhat in the 1990s.

Public relations questions raised because of the use of growth promotants and other feed additives top current issues pressing participants in the prepared feed industry. Not only do firms face increasing consumer concern stemming from the influence of feed additives on final products, such as beef, pork, milk, and poultry, but they also face various liability questions and new government regulations. Like their cohorts in the food industry, some of the things that feed labeling regulations add are requirements for tracking inputs, costs, public scrutiny, and new liability questions. Expanding trade in feeds and feedstuffs, especially with Mexico, is envisioned for the more aggressive companies following NAFTA. Corn and corn product derivatives may be stimulated by increased livestock finishing and poulty production in Mexico. However, the most potent recent issue touched off because of structural changes in the feed industry has been a price-fixing/anti-trust probe by the U.S. Justice Department. The outcome of this anti-trust investigation may have far-reaching implications for the industries associated with feeds, especially the four feed companies under investigation since mid-1995.

Conclusions

This book reports significant restructuring in the grain marketing system over the past two and a half decades. Restructuring has occurred in origination, transporting, and processing of grains and oilseeds. Firms have become larger and fewer in number. Although control has become more concentrated, the location of processing has become more decentralized. In several cases, the structural changes have included increased integration in the system from procurement through processing. Joint ventures have become a feature of mergers and consolidation.

As firms become fewer and economic power became more concentrated, the focus may shift to concern over conduct and performance of the system. The recent anti-trust action that confirmed price-fixing and market sharing indicates these issues may represent a swing back toward more focus on traditional issues of market power and its abuses. The trend to joint ventures, especially as investor-owned firms enter into arrangements with producer-owned cooperatives, may raise questions of whose interests are being served.

Developments in futures trading dramatize the problems of regulation and oversight by the appropriate agency. These developments also suggest a need for greater educational efforts by unbiased providers, which will increase the understanding by producers and local grain buyers of the many risk management alternatives available and promoted by various participants.

This chapter has summarized many of the major changes in the grain marketing system which have occurred over the past 25 years and has identified many of the drivers of those changes. A look to the future indicates many changes may yet be forthcoming which will affect marketing structure.

Continued changes in consumer preferences, demographics and life style indicate further processing of products will continue outside the home toward greater consumer orientation. These changes will likely lead to greater control over vertical food production systems with tighter links between the participants in the food production, processing, and distribution chain. There will likely be greater use of contracts between producers and processors which will require specific quality, timing of delivery, and varieties of products. The trend toward identification of niche markets will likely continue with less emphasis on homogeneous bulk commodities.

International competition in further processed food products has been increasing and will likely grow in importance. Over the last five or six years, U.S. exports of processed foods, fruits and vegetables, and meats and poultry products have increased more rapidly than traditional commodity exports. This trend will likely continue with implications for traditional marketing firms.

As price supports and supply management have become less of a national policy issue with passage of the 1996 Federal Agricultural Improvement and Reform Act, crop insurance and other forms of safety nets are becoming more of a policy issue with the role of government being an important part of the discussion. Other policy issues include U.S. sugar policy and tobacco production and regulation of the industry. Environmental issues and evolving policy actions will likely impact production and the marketing, processing, and distribution systems.